THE
NUCLEAR
ENERGY
OPTION

An Alternative for the 90s

THE NUCLEAR ENERGY OPTION

An Alternative for the 90s

Bernard L. Cohen

PLENUM PRESS • NEW YORK AND LONDON

TK
7145
C592
1990

Library of Congress Cataloging-in-Publication Data

Cohen, Bernard Leonard, 1924-
   The nuclear energy option : an alternative for the 90s / Bernard
L. Cohen.
      p.   cm.
   "Some chapters are based in part on the author's Before it's too
late"--T.p. verso.
   Includes bibliographical references and index.
   ISBN 0-306-43567-5
   1. Nuclear industry--Government policy--United States. 2. Nuclear
power plants--Government policy--United States. 3. Nuclear weapons-
-Government policy--United States. I. Title.
HD9698.U52C592 1990
333.792'4'0973--dc20   90-41744
                                                               CIP

ISBN 0-306-43567-5

Some chapters are based in part on the author's *Before It's Too Late,*
published by Plenum Press in 1983.

© 1990 Bernard L. Cohen
Plenum Press is a Division of
Plenum Publishing Corporation
233 Spring Street, New York, N.Y. 10013

Printed in the United States of America

To my grandchildren:
Shari, Aren, Lauren, Elisabeth, Carolyn,
Mollie Jane, David, and Emma Grace

CONTENTS

Chapter 1 / NUCLEAR POWER — ACT II

In the mid-1980s, nuclear power seemed to be an idea whose time had come and passed. The public seemed to have rejected it because of fear of radiation. The Three Mile Island accident was still fresh in their minds, with annual reminders from the news media on each anniversary. The Chernobyl accident in the Soviet Union in April 1986 reinforced the fears, and gave them an international dimension. Newspapers and television, on several occasions, reported stories about substandard equipment and personnel performance at nuclear power plants.

Newly completed plants were found to have been very costly, making nuclear power more expensive than electricity from coal-burning plants for the first time in 20 years. Who needed them anyhow? We already had an excess of electricity-generating capacity.

As we enter the 1990s, however, many things have changed. Environmental concerns have shifted dramatically into other areas like global warming due to the greenhouse effect, ecological destruction by acid rain, air pollution of all types but particularly from coal burning, and chemicals of various sorts, from insecticides to food additives. A succession of record-breaking warm years in the 1980s convinced many that the greenhouse effect was not just a scientific theory but might already be responsible for droughts and other agricultural disasters. Some scientists are predicting that these problems will get worse at an increasingly rapid rate in the years ahead. The president, Congress, and other politicians seem to be battling for positions of leadership in programs designed to stem the tide of the greenhouse effect. The U.S. Environmental Protection Agency is giving the matter a high priority. Nuclear plants produce no greenhouse gases, while the coal- and oil-burning plants they replace are the most important contributors.

The acid rain problem is getting increasing attention and is generating strong anti-U.S. resentment in Canada. It is also stirring up trouble in Scandinavia and Germany. Coal-burning power plants are among the worst offenders in causing acid rain, while nuclear plants avoid that problem completely.

After a decade of relative quiescence, air pollution is being reborn as a top-ranked environmental issue. The Bush Administration's Clean Air Act would require that sulfur dioxide emissions from coal-burning plants be cut in half by the end of the century, but that just scratches the surface of the problem. Nuclear plants, of course, emit no sulfur dioxide or other chemical pollutants.

The public's greatest phobias seem to have shifted away from radiation to chemicals. We have had highly publicized scares about Alar in apples and cyanide in Chilean grapes. Dioxin, PCB, EDB, and chemicals with longer names have become household words, while we hardly ever hear about plutonium anymore. The media have been carrying fewer stories about radiation and even these get little attention. In a continuing series of national polls by Cambridge Reports, the percentages of those questioned who had recently heard news about nuclear energy and who interpreted it as unfavorable shifted from about 62% and 42%, respectively, in 1983-1985 to 50% and 25% in 1989. A 1988 Roper poll found that the U.S. public considered it no more dangerous to live near a nuclear plant than a chemical manufacturing plant.

In the late 1980s, the American public learned that the radioactive gas radon was invading their homes, exposing them to many hundreds of times more radiation than they could ever expect to get from nuclear power. In fact, in some homes it was thousands or even tens of thousands of times more. But still only about 2% of the American public bothered even to test for it (at a cost of about $12), although their exposure can easily be drastically reduced. The public has perhaps grown tired of being frightened about radiation. Maybe they have caught on to the fact that after all the scare stories, there have been no dead bodies, and not even any injuries to the public. There must be a limit to how often the cry "Wolf" will be heeded.

The dramatic oil spill by an Exxon tanker near Valdez, Alaska, and the ensuing long and expensive clean-up drew constant attention to one of the environmental problems associated with a major competitor of nuclear power—oil burning. The use of oil to generate electricity is rising rapidly, and there is every reason to believe its rise will accelerate in the 1990s. A single nuclear plant can replace the oil carried by that Exxon tanker every 6 weeks.

Of course, oil spills are not the biggest problem resulting from our heavy dependence on oil. Growing concern is arising about the imbalance between our imports and exports that is gravely threatening our national economy. Imported oil is the principal villain in this matter, and the public recognizes that fact.

Lots of publicity surrounded the activities of U.S. warships in the Persian Gulf to protect oil supplies during the latter stages of the Iran-Iraq war. American lives were threatened, which set up a situation that could have led to grave consequences. Such perils are part of the price we were paying for our heavy reliance on imported oil.

Last, but far from least, of the new developments is a growing need for more electricity-generating capacity. Our expanding population and production output requires ever increasing amounts of energy, and during the decade of the 1980s, electricity's share of our total energy supply increased dramatically. There is every reason to expect this increase to continue. Serious shortages are sure to develop.

In some sections of the country, the pinch is already hurting. Brownouts (reductions in voltage which cause lights to dim, motors to turn slower, etc.) have already occurred in New England, Pennsylvania, Maryland, New Jersey, Virginia, North and South Carolina, and Chicago. Utilities in New York State have had to appeal to the public to reduce use of lighting and air conditioning, and to postpone use of dishwashers, clothes driers, and ovens. The Bonneville Power Administration in the Pacific Northwest has restricted its sales of power to Southern California. A lead story in *Fortune Magazine* (June 1989) was titled "Get Ready for Power Brownouts." A Standard and Poor publication stated "Electricity is becoming a scarce resource. . . . Power shortages in the Northeast threaten to derail the region's strong economic growth." Wall Street brokers are recommending stocks of electrical equipment suppliers because they predict a big surge in new power plant construction. Clearly, our days of excess electricity-generating capacity are at an end.

There was a prime-time TV special report on how well nuclear power is serving France, where 70% of the electricity is nuclear. The rest of the world has continued to expand its nuclear power capacity, while we have been standing still. The United States, which pioneered the development of nuclear power and provided it to the world, now ranks behind more than a dozen other nations in percentage of electricity derived from that technology. Our strongest competitor, Japan, was heavily burdened by memories of Hiroshima and Nagasaki and hence got a very late start in nuclear power, but has now far surpassed us and continues to accelerate its development.

We have well over a hundred nuclear power plants operating in the United States, and they have been steadily improving. The frequency of reactor shutdowns by safety systems has been substantially reduced. There has been little bad publicity about events relevant to nuclear safety for the past few years.

The nuclear industry has been developing a new generation of reactors that are cheaper and safer than those currently in use. They should make the concept of a reactor meltdown obsolete and make nuclear power substantially cheaper than electricity from any other source. US News and World Report featured an article on these reactors entitled "Nuclear Power, Act II." Other newspapers and magazines have carried similar stories.

All of these developments have not been lost on the American public. Attitudes toward nuclear power have been changing. A variety of public opinion polls has shown that the public is now ready to accept a resurgence of nuclear power, and indeed expects it. The stage truly seems to be set for "Nuclear Power: Act II."

Still, a decision to commit something like a hundred billion dollars to provide a substantial fraction of our nation's energy supply for the next half century through a particular technology should not be based on current whims and perceptions. These are certainly important, but they have changed in the past and can easily change in the future. The more vital questions deal with the soundness of the technology. How sound is nuclear power from the standpoints of public health and safety? of technical capability and performance? of economic viability? The American public must be educated on these matters. Its perceptions must be based on solid information to keep them from shifting unpredictably. Cycles of public acceptance and rejection of a technology designed for many decades of service are extremely expensive. We cannot afford them.

The purpose of this book is to provide this information. Most of it is scientific, obtained from sources that are generally accepted in the scientific community. Some readers may be surprised to learn that nearly all the important facts on these issues are generally accepted (within a degree of uncertainty small enough for the differences of opinion to be of no concern to the public). In the past, the media has often given the impression that there are large and important areas of disagreement within the scientific community on these matters. Actually, in spite of such attempts to dramatize it, long-standing controversy is rather rare in science. This is not to say that different scientists don't initially have different ideas on an issue, but rather that there are universally accepted ways of settling disagreements, so they don't persist for long.

Let us see how this system works. The basic instrument for scientific communication is the scientific literature, which consists principally of many hundreds of periodicals, each covering a specialized area of science. In them, scientists present research results to their colleagues in sufficient quantitative detail to allow them to be thoroughly understood and checked. They also contain critiques from researchers who may disagree with the procedures used, and replies to these critiques from the original authors, although only a few percent of the papers published are sufficiently controversial to draw such criticism. The whole system is set up to maximize exchange of information and give a full airing of the facts. On the other hand, most people would not consider the scientific literature to be interesting reading. It is written by research scientists—not usually skillful writers—to be read by other scientists in the same field. It makes extensive use of mathematics and specialized vocabulary, with little attention to techniques for holding the reader's attention. That function is provided by the scientist's professional need to obtain the information.

In addition to communication through the literature, specialists in each field frequently get together at meetings where there is ample opportunity for airing out disagreements before an audience of scientific peers. After these discussions, participants and third parties often return to their laboratories to do further measurements or calculations, developing further evidence. In most cases, controversies are thereby settled in a matter of months, leading to a consensus with which over 90% of those involved would agree. Where scientific questions have an impact on public policy, there is an additional mechanism. The National Academy of Sciences and similar national and international agencies assemble committees of distinguished scientists specializing in the field to develop and document a consensus. Only very rarely do these committees have a minority report, and then it's from a very tiny minority. The committees' conclusions are generally accepted by scientists and government agencies all over the world.

Since controversy is the rule rather than the exception in human affairs, many find it difficult to believe that science is so different in this respect. The reason for the difference is that science is based largely on quantities that can be measured and calculated, and these measurements and calculations can be repeated and checked by doubters, with a very heavy professional penalty to be paid by anyone reporting erroneous results. This ability to rapidly resolve controversy has been one of the most important elements in the great success of science, a success that during this century has increased our life expectancy by 25 years, and improved our standard of living immeasurably.

A minority of the material to be discussed in this book is nonscientific, and even political, covering areas in which I have no professional expertise. For this I depend on my reading of the general literature and attendance at lectures over many years. I cannot vouch for this material's reliability, but fortunately it is largely noncontroversial.

When specialists present information to the public, they can easily convey false impressions without falsifying facts by merely selecting the facts they present. It is my pledge not to do this. I will do my utmost not only to present correct information but to present it in a way that gives the correct impression and perspective. To do otherwise would seriously damage my credibility in the scientific community and thereby depreciate the value of my research, which is largely what I live for.

Since your faith in this pledge may depend on what you know about the author, I offer the following personal information. I am a 65-year-old, long-tenured professor of physics and radiation health at the University of Pittsburgh. I have never been employed by the nuclear industry except as a very occasional consultant, and I discontinued those relationships several years ago. My job security and salary are in no way dependent on the health of the nuclear industry. I have no long-standing emotional ties to nuclear power, not having participated in its development. My professional involvement with nuclear energy began only when the 1973 oil embargo stimulated me to look into our national energy problems. I have four children and eight grandchildren; my principal concern in life is to increase the chances for them and all of our younger citizens to live healthy, prosperous lives in a peaceful world.

To those who question my selection of topics or my treatment of them in this book, I invite personal correspondence or telephone calls to discuss these questions. I feel confident that through such means I can convince any reasonable person that the viewpoints expressed are correct and sufficiently complete to give the proper impressions and perspective.

Other scientists have written books on nuclear energy painting a very different picture from the one I present. Ernest Sternglass,[1] John Gofman,[2] and Helen Caldicott[3] are the names with which I am familiar. Their basic claim is that radiation is far more dangerous than estimates by the scientific Establishment would lead us to believe it is. This is a scientific question which will be discussed in some detail in Chapter 5, but the ultimate judgment is surely best made by the community of radiation health scientists. A poll of that community (see Chapter 5) shows that the scientific works of these three scientists have very low credibility among their colleagues. Their ideas on the dangers of radiation have been unanimously rejected by various committees of eminent scientists assembled to make judgments on those questions. These committees represent what might be called "the Establishment" in radiation health science; the poll shows that they have very high credibility within the involved scientific community. In a secret ballot, less than 1% gave these Establishment groups a credibility rating below 50 on a scale of 0-100, whereas 83% gave the three above-mentioned authors a credibility rating in that low range. The average credibility rating of these Establishment groups was 84, whereas less than 3% of respondents rated the three authors that high.

The positions presented in this book are those of the Establishment. In comparing this book with those of Sternglass, Gofman, and Caldicott, you therefore should not consider it as my word against that of those three authors, but rather as their positions versus the positions of the Establishment, strongly supported by a poll taken anonymously of the involved scientific community.

Another approach for distinguishing between this book and theirs is the degree to which the authors are actively engaged in scientific research. The Institute for Scientific Information, based in Philadelphia, keeps records on all papers in scientific journals and publishes listings periodically in its *Science Citation Index—Sourcebook*. The number of publications they list for the various authors, including myself, are:

Author	1975-1979	1980-1984	1985-1988
H. Caldicott	3	2	1
J. W. Gofman	2	3	0
E. J. Sternglass	5	9	4
B. L. Cohen	65	53	37

It might be noted that most of the papers by Gofman were his replies to critiques of his work, and most of the entries for Sternglass were papers at meetings which are not refereed (the others were on topics unrelated to nuclear power). Only a small fraction of my papers are in these categories. Since 1982 I have been directing a large experimental project involving measurements of radon levels in 350,000 houses. This has reduced the number of papers I have produced.

Of course, many books about nuclear power have been written by professional writers and other nonscientists. They are full of stories that make nuclear power seem dangerous. Stories are useful principally to maintain reader interest, but they don't prove anything. In order to make a judgment on the hazards of nuclear power, it is necessary to quantify the number of deaths (or other health impacts) it has caused, or can be expected to cause, and compare this with similar estimates for other technologies. This is something that the books by nonscientists never do. They sometimes quantify what "might" happen, but never quantify the probability that it will happen. If we are to be guided by what might happen, I could easily concoct scenarios for any technology that would result in more devastation and death than any of their stories about nuclear power.

A book of this type must often get into discussions of scientific details. Every effort has been made to keep them as readable as possible for the layperson. The more technical details have been put into Appendixes at the end of the book. These can be ignored by readers with less interest in details. For readers with more interest in these, references are given which can be used as starting points for further reading. Personal inquiries about further information or references are always welcome.

Each chapter is broken up into sections. If a reader is not interested in the subject of a particular section or finds it to be too technical, it can usually be skipped over without loss of continuity.

With these preliminaries out of the way, let us begin by asking, "Do we need more power plants for generating electricity?"

Chapter 2 / DO WE NEED MORE POWER PLANTS?

Power plants for generating electricity cost billions of dollars, which is very expensive by any standard, and a plant's cost must be borne by the utility that will sell its output. A utility will therefore not build a power plant unless it is needed and will yield an eventual profit. How does a utility decide to build such a plant?

Over the first 70 years of this century, the consumption of electricity grew at a steady rate of 7% per year, doubling every 10 years. In the early 1970s, it was therefore natural for utilities to assume that this trend would continue, so they planned for the construction of new power plants accordingly. Then, suddenly, in the wake of the 1973-74 energy crisis, the growth in electricity consumption slowed down, as conserving energy became the order of the day. The inflation and economic downturns that followed continued to have a depressing effect. Between 1973 and 1982, average growth in electric power consumption was only a little over 1% per year. That meant that the construction programs for new plants were leading to a gross excess of electrical capacity. Utilities scurried to cancel construction projects, both nuclear and coal. But in many cases, cancellation costs were so high that they decided to finish construction. This led to a substantial excess of generating capacity in the early 1980s. Consequently, there have been very few new construction starts since that time.

Beginning in 1982, a long period of economic prosperity developed, causing growth in electricity consumption. It increased at an average rate of 4.5% per year between 1982 and 1988, a total of 27% over the 6-year period. In 1989 it increased by another 2.6%. These increases have eliminated nearly all of the excess capacity and threaten imminent shortages.

With most commodities, production need not follow the ups and downs in consumption because excess production can be stored for later use, but with electricity that is not normally possible. It ordinarily must be generated as it is used. There is an alternative of pumping water into a reservoir high up on a mountain when excess electricity is available, as in the middle of the night, and then have it flow down, producing hydroelectric power as needed. But most proposals of this type have never "gotten off the ground" because of costs and opposition by environmental groups on the basis of land use. Consequently, a utility must have enough generating capacity available to meet peak loads. A hot summer afternoon, when air conditioners are going full blast, is usually the critical time, but cold, dark winter days are also a challenge.

In a crunch, utilities can, to some extent, buy power from one another through interconnecting lines. But the capacity of these interconnections is limited, and there are problems involved in increasing it. A 10-mile interconnecting link around Washington, D.C., has been held up for 13 years by environmental opposition to unsightly power lines and the land they require for rights of way. Furthermore, neighboring utilities may not have excess power to sell, and importing power from far away is both inefficient and expensive.

For a typical utility, an average of 14% of its generating capacity is shut down for maintenance and repair at any given time, but this figure is subject to fluctuations. It is impossible to be sure that a large plant will not have to be shut down unexpectedly. Therefore, a reserve capacity of 20% above normally expected requirements is considered to be reasonable, with anything below 17.5% somewhat precarious for reliable service. Since efficient power plants take at least 4 years to construct, we now can predict how many will be available in 1993. If our use of electricity expands at a rate of only 2.5% per year—recall that it has been expanding at nearly twice that rate since 1982—much of the East coast will have less than 15% reserve capacity by

1993. Large areas in the East, Southeast, and the Great Lakes region will have less than 17.5%, and the great majority of the nation east of the Rocky Mountains will have less than 20%. The recent trend in construction has been to increase power plant capacity by only about 1% per year, which means that, unless the expansion of electricity consumption slows down dramatically, things will continue to get worse beyond 1993.

The electricity shortage is now most severe in Florida because that state has been growing so rapidly, twice as fast as the national average. On Christmas day of 1989, peak demand reached 33.8 million kW, whereas the amount the utilities could produce was only 24.8 million kW, and the quantity that could be brought in from out of state was only 3.4 million kW, leaving a shortage of 5.6 million kW, or 16%.

What happens when a utility's capacity is exceeded?[1] If it is exceeded by only a few percent, the voltage is reduced. This causes lights to dim, and is called a *brownout*. Appeals to the public to conserve electricity follow. "Nonessential" users are asked to shut down for a day. That raises the disturbing question of which users are nonessential, and who decides. In the hot summer of 1988, New England went through 10 of these brownouts, including one that shut down Harvard University and Boston's City Hall for a day. It was only the third time in its 353-year history that Harvard had closed down. Other brownouts were mentioned in Chapter 1.

If the capacity is exceeded by 15% or more, the utility must resort to rolling *blackouts,* cutting off service to various neighborhoods for a few hours per day. During the cold Christmas season of 1989, there were rolling blackouts in Tampa, Jacksonville, and several other places in Florida, and the first blackout in Texas history hit Houston. Blackouts lead to all sorts of inconveniences, but refrigeration problems are especially severe. Without it, fish and dairy products are the first to spoil, with meat following close behind. In the summer of 1978, the utility serving Key West, Florida, had to resort to 15% rolling blackouts for 26 days to repair equipment, and citizens were throwing spoiled meat through its office windows. The public does not readily accept electricity shortages.

WHAT TYPES OF PLANTS SHOULD WE BUILD?

It seems obvious from the above discussion that our nation needs lots of new plants for generating electricity. The next issue to be confronted is what type should be built. The most efficient way to produce electricity is with turbines driven by an externally provided fluid, usually steam or water. The turbine itself is a machine which converts the motion of this fluid into the turning of a shaft. This turning shaft is then coupled to the shaft of a generator, somewhat like the generator or alternator in an automobile but enormously larger, which converts its energy into electricity.

The turbine may be driven by falling water, a takeoff on the familiar water wheel but very much larger and more highly engineered so as to convert nearly all of the energy of the falling water into electrical energy. This hydroelectric power is relatively cheap and has been important historically. Large plants harnessing the energy of Niagara Falls, the Tennessee River and its tributaries, the Columbia River and its tributaries, with its showpiece Grand Coulee Dam, and the Colorado River, including Hoover Dam, are familiar to tourists and have played very important roles in the economic development of the surrounding areas. Norway derives most of its electric power from this source, as does the Canadian province of Quebec and several other areas around the world. But sites for generating hydroelectric power must be provided by nature, and in the United States nearly all of the more favorable sites nature has provided are already being used. There have been new projects for harnessing the energy in the flow of rivers, but these give relatively little electric power and cause serious fish kills, which lead to well-justified objections by environmental groups. Hydroelectric power is therefore not an important option for the new plants that are needed.

Turbines can also be driven by wind. There has been a lot of activity in this area (see Chapter 14), but there are lots of problems. Thousands of large windmills, each the height of a 20-story building, would be needed to substitute for a single conventional power plant. There would be serious environmental effects, including noise and interference with birds. But most important, there is the very sticky question of what to do when the wind isn't blowing. Using batteries to store the electricity is far too expensive.

In the great majority of large power plants in the United States and in nearly every other nation, the turbines are driven by the pressure of expanding steam. When the steam is heated to very high temperatures (600°F or higher), it contains a lot of energy, and highly engineered turbine-generators are available for converting a large fraction of this energy into electricity. This steam is produced by boiling water and superheating the vapor, which requires a great deal of heat. The question is how will this heat be provided?

It can be produced by burning any fuel, but since coal is the cheapest fuel, most electricity in the United States is generated by coal burning. We have enough coal in the United States for hundreds of years to come, and it is relatively low in cost. The problems with the use of coal are largely environmental. They will be discussed in the next chapter.

The other major fossil fuels, oil and natural gas, are much more expensive in most parts of our country, but they are used to some extent. However, our second most prevalent way of producing this steam is with nuclear reactors, a technology referred to as nuclear power.

Nuclear power provides a quarter of all electricity in industrialized countries.[2] In 1988 it provided 70% of the electricity used in France, 66% in Belgium, 49% in Hungary, 47% in Sweden and Korea, 41% in Taiwan, 37% in Switzerland, 36% in Spain, Finland, and Bulgaria, 34% in West Germany, 28% in Japan (increasing rapidly), 27% in Czechoslovakia, 19% in the United States and United Kingdom, and 16% in Canada. Nuclear power plants are now operating in 27 different countries and are under construction in 5 others. They are clearly a major contender.

While power plants utilizing steam are the preferred technology in the great majority of situations, they have one serious drawback — they require at least 4 or 5 years to construct and put into operation. If a crunch develops, requiring that new generating capacity be made available quickly, the only real option is internal combustion turbines. In them, the turbine is driven by the hot gases produced by the burning of oil or natural gas, similar to the way in which pistons are driven in an automobile engine.

Internal combustion turbines can be purchased, installed, and put into operation in 2 or 3 years. As a result of the crunch which is developing now, purchases of these machines are escalating sharply. Orders for new internal combustion turbines by U.S. utilities, in millions of kilowatts capacity, were 3.1 in 1987, 4.4 in 1988, and 5.3 in 1989. They are relatively cheap to purchase and install, which makes them attractive to utilities. Their principal drawbacks are that they are inefficient and their fuel — oil and/or natural gas — is expensive. These drawbacks are not a problem for utilities, since they have no difficulty in convincing the public utility commissions which regulate them to pass the fuel charges directly on to consumers. Internal combustion turbines are therefore easy for utilities to accept, but they increase electricity costs to consumers. They also increase our oil imports. In fact, these machines themselves are mostly imported, contributing to our balance-of-payments problems. There are many other important reasons why we should avoid expanding our use of oil.

Why Not Use Oil?

Oil is the principal fuel used in the United States and throughout the world. It is essentially the only fuel that runs our automobiles, trucks, airplanes, and ships, and is the dominant fuel for buses and railroads. Without it, our transportation system would grind to a screeching halt. Natural gas is the principal fuel for heating buildings, but oil also carries an appreciable part of that load.

In addition, oil has many vital uses other than as a fuel. Plastics, organic chemicals, asphalt, waxes, and essentially all lubricants are derived from it. Tremendous adjustments would be required if society had to forego use of almost any one of these items, let alone all of them. For example, nearly all industrial processes depend on organic chemicals, and many medicines, pesticides, paints, and the like, are manufactured from them.

The world's oil supplies are quite limited, hardly enough to get us halfway through the next century if present usage trends continue. In this situation it hardly makes sense to burn oil when other fuels can do the job just as well and less expensively.

But for the near-term future, at least, the *distribution* of oil throughout the world is a more serious problem than the total world supply. The nations that need it are not the nations that have it. Western Europe has very little and Japan has none. The United States has a reasonable amount, but not nearly enough to satisfy our needs. At the time of the oil crisis in 1974, we were importing 36% of our oil. By 1984, with the help of Alaskan oil and vigorous conservation measures, this was reduced to 29%. For example, in 1974 about 17% of our electricity was generated from oil, but by 1986 this was reduced to 5% by converting oil-burning power plants to coal. Laws required that automobiles be made smaller and more fuel efficient, and the national speed limit was reduced from 65 to 55 miles per hour, giving a substantial fuel saving. Thermostats were lowered in homes. It seemed like the situation was under control.

But in recent years, more and more oil has been used to generate electricity—23 million gallons per day in 1987, 27 in 1988, and 33 in the first half of 1989. A Georgetown University study projects that it will reach 84 million gallons per day by 1995. Other uses of oil have also been expanding. By 1988 we were importing over 40% of our oil, and in 1989 it was 46%; the trend is clearly climbing upward.[3] At the same time, U.S. production is falling off; in 1989 it decreased by 5.7%. The prospects for the future look bleak. Alaska is now producing about 9% of our oil, but the Prudhoe Bay reservoir is being depleted and production will soon be falling off sharply. At the same time, our oil consumption is increasing by about 1% per year.

As a result, OPEC's control of the world oil market is being reestablished. The sharp decline in oil prices in the mid-1980s occurred because demand was down to the point where OPEC was producing at only 60-70% of capacity. By 1989 it was up to 80%, as compared to about 85% when oil prices peaked in 1979-80. The honeymoon on oil prices may rapidly draw to a close unless the OPEC countries decide to keep it going, further cementing their strong position by encouraging world oil consumption to continue its rapid growth. That would only postpone the day of reckoning, and worsen its effects.

In 1989, the United States spent over $40 billion on imported oil, which represented about 40% of our trade deficit. If the price returns to its 1980 level, this cost will double. In fact, most estimates predict that oil imports will be costing us more than $100 billion per year before the end of the century. The impacts this will have on our economy could be severe.

In addition to the economic difficulties, there is good reason to worry about political issues. The fall of the Shah of Iran doubled world oil prices. Many of the OPEC countries are also politically unstable. We are highly vulnerable to cutoff of our oil supplies, or to unreasonably sharp price increases.

There is a real risk that oil politics may lead to war. In 1988 our navy was involved in serious shooting incidents in the Persian Gulf. If Iran and Iraq had not been exhausted by years of war, things could have been much worse. When one thinks about scenarios that could lead to a worldwide conflict, many would involve fighting over Middle East oil.

Oil also has its environmental problems, to be discussed in the next chapter. For all of these reasons, students and practitioners of energy strategies unanimously agree that measures which increase our use of oil must be avoided as much as possible.

Problems with the Use of Natural Gas

The problems with the use of natural gas for generating electricity largely parallel those for oil. Gas also has many other uses for which it is uniquely suited, like heating buildings and furnaces in industry (e.g., melting steel or glass). It can readily be used as a substitute for gasoline to drive automobiles; it is now widely utilized in Canada for that purpose. It is the principal alternative to oil for manufacture of plastics and organic chemicals, and in many cases it is more valuable for that purpose because the required conversion processes are simpler and cheaper.

As in the case of oil, world supplies are hardly sufficient for the next half century if present usage patterns persist. Natural gas has an advantage over oil in being largely of domestic origin (with important contributions from Canada and Mexico), but in most areas it is more expensive than the type of oil used to generate electricity. In many areas, this cost would be prohibitive.

Natural gas is the principal fuel for generating electricity in Texas and surrounding states where it is cheap, but nationally it provides only 9% of our electric power. This situation has been largely controlled by price considerations and uncertainty about availability of future supplies. Gas producers are not willing to sign long-term contracts, and it is not difficult to understand why a utility hesitates to build a billion dollar plant without such a contract.

While natural gas seems like a better choice than oil for future power plants, it is far from ideal. It is limited in supply, very much needed for other purposes, and generally expensive.

ELECTRICITY AS A SUBSTITUTE FOR OIL AND GAS

While we have been discussing the problems of using oil or gas to generate electricity, most of those involved in planning energy strategies put more emphasis on the opposite, using electricity generated from plentiful coal and nuclear fuels to substitute for oil and natural gas. This substitution has been going on continuously for some time and has played a key role in reducing our usage of oil and gas. Between 1973 and 1988, our use of electricity increased by 50%, while our use of all other types of energy decreased by 5%. During this period, the percentage of our energy used in the form of electricity increased from 27% to 37%. It is expected to exceed 50% early in the next century.

For most applications, electricity is more convenient than burning fuels. It is cleaner and very much easier to control. A flick of a finger turns it on or off, and a twist of the wrist increases or decreases the flow. It is especially amenable to computer control, since computer inputs and outputs are electrical. It is easier to maintain, more flexible, and generally safer. These are the reasons for its rapid growth, and the reasons for believing the trend will continue.

When gas and oil become scarce and their prices rise, this trend will undoubtedly accelerate. For example, electric cars and railroads can save vast quantities of oil, and electrically driven heat pumps are already displacing natural gas for heating buildings. These shifts will greatly enhance the need for new power plants in the next century.

COAL OR NUCLEAR?

By now, hopefully, the reader is convinced that we are going to need lots of new power plants in the very near future and that most of them should not be burners of oil or gas. The only real alternatives for the great majority of these are coal or nuclear plants. (In some areas remote from coal resources, like New England and most of our Pacific and Atlantic coasts, the only alternatives are oil and nuclear). The two should be roughly cost competitive. But the decision will largely be made on the basis of environmental impacts. That is what the rest of this book is all about.

Chapter 3 / ENVIRONMENTAL PROBLEMS WITH COAL, OIL, AND GAS

Over the past quarter century, we have come to realize that there is more to life than material goods and services, that "some of the best things in life are free." The pleasure we derive from breathing fresh air, drinking pure water, and enjoying the beauty that nature has provided is priceless and must not be sacrificed. Moreover, losing them will lead directly or indirectly to incalculable economic losses. We have come to appreciate the importance of our environment.

Much has been said and written about environmental problems with nuclear power, and they will be discussed at great length in this book. But in this chapter, we consider the wide variety of environmental problems in burning fossil fuels—coal, oil, and gas. They probably exceed those of any other human activity. The ones that have received the most publicity in recent years have been the "greenhouse effect," which is changing the Earth's climate; acid rain, which is destroying forests and killing fish; and air pollution, which is killing tens of thousands of American citizens every

19

year, while making tens of millions ill and degrading our quality of life in other ways. We will discuss each of these in turn, and then summarize some of the other problems that have drawn lesser attention. But first we must begin with some basics.

Coal, oil, and gas consist largely of carbon and hydrogen. The process that we call "burning" actually is chemical reactions with oxygen in the air. For the most part, the carbon combines with oxygen to form carbon dioxide (CO_2), and the hydrogen combines with oxygen to form water vapor (H_2O). In both of these chemical reactions a substantial amount of energy is released as heat. Since heat is what is needed to instigate these chemical reactions, we have a *chain reaction*: reactions cause heat, which causes reactions, which cause heat, and so on. Once started the process continues until nearly all of the fuel has gone through the process (i.e., burned), or until something is done to stop it. Of course, the reason for arranging all this is to derive the heat.

The carbon dioxide that is released is the cause of the greenhouse effect we will be discussing. A large coal-burning plant annually burns 3 million tons of coal to produce 11 million tons of carbon dioxide. The water vapor release presents no problems, since the amount in the atmosphere is determined by evaporation from the oceans—if more is produced by burning, that much less will be evaporated from the seas.

THE GREENHOUSE EFFECT

Electromagnetic radiation is an exceedingly important physical phenomenon that takes various forms depending on its wavelength. (The concept of *wavelength* is most easily understood for water waves, where it is the distance between successive crests. For electromagnetic waves it is the distance between successive peaks of the electric and magnetic fields.) Ordinary radios use the longest wavelengths of interest, 200 to 600 meters for AM stations. FM radio and television use wavelengths from a few meters (UHF) to less than 1 meter (VHF). Microwaves, familiar from the ovens that employ them, and radar, which plays a vital role in military applications and is also used by the police to catch speeders, have shorter wavelengths, in the range of centimeters to millimeters. Visible light is electromagnetic waves with much shorter wavelengths, ranging from 0.0004 millimeters for purple to 0.00055 millimeters for green to 0.0007 millimeters for red. Wavelengths between where visibility ends (0.0008 millimeters) and the microwave region begins (0.1 millimeters) are called *infrared*.

Every object in the universe constantly emits electromagnetic radiation, and absorbs (or reflects) that which impinges on it. According to the laws of physics, the wavelength of the emitted radiation decreases inversely as the temperature increases. (Here we use absolute temperature, which is Fahrenheit temperature plus 460 degrees.) For example, the filament of a light bulb, which is typically at 6,000 degrees absolute, emits visible (yellow) light, while our bodies, which are normally at 559 degrees absolute (460 + 98.6), emit radiation of about 11 times longer wavelength, which is in the infrared region and therefore not visible — that's why we don't glow in the dark. More applicable to our discussion is the surface of the sun, which is at 11,000 degrees and therefore emits visible light as we can plainly see, and the surface of the Earth which, at about 520 degrees, emits infrared, as we know from the fact that it is not visible on a dark night.

The rate at which an object emits radiation energy increases very rapidly with increasing temperature (doubling the absolute temperature increases the radiation 16-fold). That's why a hand placed near a light bulb is heated much more by radiation from the filament when it is hot (light on) than when it is cool (light off). Now let us consider a bare object out in space, such as our moon. It receives and absorbs radiation from the sun, which increases its temperature, and this increased temperature causes it to emit more radiation. Through this process it comes to an equilibrium temperature, where the amount of radiation it emits is just equal to the amount it receives from the sun. That determines the average temperature of the moon. If this were the whole story, our Earth would be 54 degrees cooler than it actually is, and nearly all land would be covered by ice.

The reason for the difference is that the Earth's atmosphere contains molecules that absorb infrared radiation. They do not absorb the visible radiation coming in from the sun, so the Earth gets its full share of that. But a fraction of the infrared emitted by the Earth is absorbed by these molecules which then reemit it, frequently back to the Earth. That is what provides the extra heating. This is also the process that warms the plants in a greenhouse — the glass roof does not absorb the visible light coming in from the sun, but the infrared radiation emitted from the plants is absorbed by the glass and much of it is radiated back to the plants. That is how the process got its name — greenhouse effect. It is also the cause of automobiles getting hot when parked in the sun; the incoming visible radiation passes through the glass windows, while the infrared emitted from the car's interior is absorbed by the glass and much of it is emitted back into the interior.

Molecules in the atmosphere that absorb infrared and thereby increase the Earth's temperature are called greenhouse gases. Carbon dioxide is an efficient greenhouse gas. The atmosphere of Venus contains vast quantities of carbon dioxide, elevating its temperature by 500 degrees over what it would be without an atmosphere. That's why no astronaut will ever be able to land on Venus. Mars, on the other hand, has no atmosphere and is at the temperature expected from the simple considerations discussed above in connection with our moon.

Our problem is that burning coal, oil, and gas produces carbon dioxide, which adds to the supply already in the atmosphere, increasing the greenhouse effect and thereby increasing the temperature of the Earth. Prior to the industrial age, the concentration of carbon dioxide in the atmosphere was less than 280 ppm (parts per million). This was determined from analyzing air bubbles, trapped hundreds or thousands of years ago in the Antarctic and Greenland ice caps. By 1958 the carbon dioxide concentration had risen to 315 ppm, and by 1986 it was 350 ppm. The average temperature of the Earth has been about 1 degree warmer in the 20th century than in the 19th century, which is close to what is expected from this carbon dioxide increase. As the rate of burning coal, oil, and gas escalates, so too does the rate of increase of carbon dioxide in the atmosphere.

Predicting the increase of temperature expected from this is very complicated, but since it is so important, a great deal of effort has gone into deriving estimates. Results are usually discussed in terms of doubling the concentration of carbon dioxide, from 350 ppm to 700 ppm. If current trends continue, this will occur during the next century, perhaps as early as 2030.[1] The direct effect of doubling the carbon dioxide in the atmosphere would be to raise the Earth's average temperature by 2.2°F. Two side effects will accentuate this temperature rise. One is that the increased temperature causes more water to evaporate from the oceans, which adds to the number of water molecules in the atmosphere; water vapor is also a greenhouse gas. The other is that there would be less ice and snow; these reflect away the visible light from the sun that would otherwise be absorbed by the Earth's surface.

There are many other factors of lesser importance that must be considered. Some of these factors tend to reduce the warming effect:

- Clouds, generated by the increased evaporation of water, intercept some of the radiation coming in from the sun and emit part of it back into outer space.
- Volcanoes inject lots of dust into the atmosphere; this dust reflects sunlight away from the Earth.
- Plankton, tiny marine organisms whose growth is accelerated by

carbon dioxide and higher temperatures, absorb carbon dioxide, thereby taking it out of circulation.
- Oceans absorb both carbon dioxide and heat.

Yet other complicating processes accentuate the warming:

- Sulfur dioxide, a pollutant we will be discussing soon, tends to cool the Earth, and in our efforts to eliminate pollution we are reducing this cooling effect.
- Thawing of permafrost, soil that has been frozen for thousands of years, releases methane (natural gas), which is a greenhouse gas.
- Bacteria in soil convert dead organic matter into carbon dioxide more rapidly as temperatures rise, thus increasing the amount of carbon dioxide in the atmosphere.

When all of these factors are taken into account as accurately as possible with our present knowledge, the best estimates are that doubling the carbon dioxide in the atmosphere will increase the average temperature by about 7°F. The uncertainty in this estimate is large; the true increase might be as little as 3° or as much as 15°F.

The importance of this greenhouse effect has become a public issue because of the recent abnormally hot summers accompanied by droughts that have severely reduced our agricultural output. Averaged over the Earth, the five warmest years in the past century have been in the 1980s, this despite the facts that the sun's energy output has been below normal and that there has been major volcanic activity, which would ordinarily reduce temperatures as explained above. Whether or not this recent abnormally warm weather is a manifestation of the increasing greenhouse effect is somewhat debatable, but unquestionably the greenhouse effect will become important sooner or later if we continue to use fossil fuels. There is, therefore, a strong consensus in both the scientific and environmentalist communities that the greenhouse effect should be given high-priority attention.

Since our emphasis here is the generation of electricity, we have concentrated our discussion on carbon dioxide from burning fossil fuels, but that is only part of the story on the greenhouse effect. Other aspects are discussed in the Chapter 3 Appendix. We now turn our attention to the consequences of this global warming.

PREDICTED CONSEQUENCES OF THE GREENHOUSE EFFECT

In December 1987, the U.S. Congress requested a report from the Environmental Protection Agency (EPA) on the health and environmental consequences expected from the greenhouse effect. The following discussion is based on that report.[2]

Agriculture is especially sensitive to climate. For example, the hot, dry summer of 1988 reduced corn yields in the Midwest by 40%. But with long-term planning, crops can be changed to compensate for climate change. Moreover, increased levels of carbon dioxide would have beneficial effects on agriculture, since carbon dioxide in the air is the principal source of material from which plants produce food.

The EPA's assessment is that the South will be hard hit, as the temperature becomes too hot for most of its crops, especially soybeans and corn. Florida is an exception, as citrus growing will be helped and tropical fruits can become a major new product. The Great Lakes region will be helped by the longer growing season. Crop yields in Minnesota will be increased 50-100%, with much lesser benefits elsewhere. Corn growing will become difficult in Illinois, but it can be replaced by sorghum. The Great Plains region will suffer the most, as it is already somewhat marginal for agriculture. Its major crops, wheat and corn, will probably have to be abandoned.

Livestock problems will increase. Heat stress will reduce breeding. Some of the livestock diseases that now plague the South will shift northward, and new tropical diseases will invade the South. Problems with agricultural pests will multiply. More pests will survive over the warmer winters, and they will breed more generations over the longer summers.

In summary, while there will be lots of disruptions and requirements for adjustment, the EPA does not expect food shortages to become critical in the United States. Problems could be more difficult in other parts of the world.

Forests will undergo some hard times. Each type of tree requires a specific climate. Thus, the growing area for each species will shift northward by 100 to 600 miles. This sounds innocuous, but adjustment periods will be difficult, with lots of die-off in the South and slow build-up in the North. Forests are constantly under stress from insects, diseases, competition with other plants, fires, wind, and the like. The added stress of changing climate is certain to cause trouble. A great deal of research and planning will be needed to cope with it.

A substantial fraction of U.S. cities are on seacoasts and thus close to sea level — Boston, New York, Baltimore, Washington, Miami, New Orleans, Houston, San Diego, Los Angeles, San Francisco, and Seattle, to name a few. If all snow and ice were to melt, sea level would rise by 270 feet, enough to inundate nearly all of these cities and vast other areas of the nation. If present trends continue, there will be a 20-foot rise in 200-500 years. A reasonable estimate for the middle of the next century is 1.5-3 feet.

A 3-foot rise would flood major areas in Boston, New York, Charleston, Miami, and especially New Orleans, and in all would reduce the land area of the United States by an area equal to that of the state of Massachusetts. Most of this land loss would be in Louisiana and Florida. The barrier islands off our coasts, including Atlantic City and Miami Beach, would be in severe trouble.

In general, each 1-foot rise in sea level moves the coastline back 50 to 100 feet in the northeast, 200 feet in the Carolinas, 200 to 400 feet in California, 100 to 1,000 feet in Florida, and a few miles in Louisiana. The flooding now caused by storms once in 100 years would be expected about every 15 years. Since hurricanes are generated in warm water (above 79 degrees), more of them would be expected.

Inland penetration of salt water would cause lots of difficulties for aquatic life. This is already reducing oyster harvests in Chesapeake Bay. New York City derives its drinking water from the Hudson River just above the present penetration of salt water; that would have to be changed. Philadelphia and several other cities have similar problems. Contamination of groundwater with salt would be a widespread problem, especially in Florida.

For valuable land, like cities, effective measures can be taken to control flooding from rising sea levels. Dikes can be built with large capacity pump-out systems to handle overflow as in Holland. This can be used to protect major U. S. cities against a 6-foot rise in sea level for $30-100 billion. New buildings can be constructed on higher ground, back from the shore. Fill can be added to raise the land used for new construction. Again there will be lots of problems, but with wise planning, they can be solved or at least delayed.

There is also a concern for wetlands, which are important for waterfowl and some types of aquatic life. It would take only a small rise in sea level to cut the total U.S. wetlands in half.

Rising ocean levels do not necessarily mean rising levels in rivers and streams. For example, the levels of the Great Lakes are predicted to fall 2-5 feet, due to the greenhouse effect reducing rainfall and increasing evaporation. This will cause problems with shipping and with water supplies for cities and towns.

Wild animals and plants must adapt to climate changes, and there are many potential difficulties. In some situations, animals can simply move, but not always. The grizzly bears in Yellowstone park would have no place to go and probably would die out. Other casualties of the greenhouse effect will probably be panthers, bald eagles, and spotted owls.

Insect plagues can be expected to cause lots of problems for trees, as will increased floods and droughts. Forest fires will occur more frequently.

Acid rain problems, to be discussed later, will become worse. It will not be an easy time for forests or the animals that inhabit them.

While all of these things are going on in the United States, the rest of the world will also be affected. When one thinks of rising sea levels, Holland comes to mind. Two-thirds of its land area is now below sea level, and even now it is protected by dikes against a 16-foot rise in sea level during storms. For $5-10 billion, Holland can be protected against a further 3-foot rise. Technology really helps in this matter. Bangladesh, with its low-technology society, is expected to suffer terribly from flooding as the sea level rises.

One would think that Canada would benefit from warmer climates, but there are many complications. With the oceans rising and the Great Lakes water levels falling, the St. Lawrence seaway will be in trouble. Southern Ontario, which has the most productive farmland in Canada, may suffer from drought as storm tracks and the rain they bring move north, and warmer temperatures cause increased evaporation. The western wheat belt will also be threatened by drought. But aside from these local problems, in general the greenhouse effect causes agriculture to move northward, and Canada can accommodate a lot of northward movement. The tree line moves north by about 35 miles for each degree Fahrenheit of global temperature rise.

Most of the above discussion is based on what will happen by the middle of the next century. But that is not the end of it. As long as we burn fossil fuels, the Earth's climate will continue to get warmer. The only solution is to strongly reduce our burning of coal, oil, and gas. Substituting nuclear energy for coal burning to generate electricity, and the substitution of electricity for oil and gas in heating buildings and to some extent in transportation, can play an important role in this process.

ACID RAIN

In addition to combining carbon and hydrogen from the fuel with oxygen from the air to produce carbon dioxide and water vapor, burning fossil fuels involves other processes. Coal and oil contain small amounts of sulfur, typically 0.5% to 3% by weight. In the combustion process, sulfur combines with oxygen in the air to produce sulfur dioxide, which is the most important contributor to acid rain. Air consists of a mixture of oxygen (20%) and nitrogen (79%), and at very high temperatures molecules of these can combine to produce nitrogen oxides, the other important cause of acid rain. Sulfur dioxide and nitrogen oxides undergo chemical reactions in the atmosphere to become sulfuric acid and nitric acid, respectively, dissolved in water droplets that eventually may fall to the ground as rain. This rain is therefore acidic.

Chemists measure acidity in terms of pH, on a scale which varies from 0 to 14. On this scale, a pH of 0 is maximum acidity, a pH of 7 is neutral, and a pH of 14 is maximum alkalinity, which is the opposite of acidity. Each unit of pH represents a factor of 10 in acidity. For example, a pH of 4 is 10 times more acidic than a pH of 5. Some examples of pH values are 1.2 for the sulfuric acid in automobile batteries, 2 for lemon juice or vinegar, 3 for apple juice, 4 for tomato juice, 5 for carrot juice, 6.3 for milk, 7.3 for blood, 8.5 for soap, and 13 for lye, which is almost a pure alkali.

We have seen that there is a substantial amount of carbon dioxide in the atmosphere. This dissolves in water droplets to form carbonic acid, familiar to us as carbonated water or soda water. As a result, "natural" rain is somewhat acidic, with a pH of about 5.6. Other natural factors, such as volcanic activity, also contribute, causing wide variations in pH. Rainfall pH as low as 4.0 has been observed even in places remote from the effects of fuel burning, like Antarctica and the Indian Ocean.

There is evidence that rain is made appreciably more acidic by the sulfuric and nitric acid from fossil fuel burning. For example, one study indicated that in the eastern United States, the pH of rain was in the range 4.1-4.5 in the 1980s versus 4.5-5.6 in the 1950s.

After the rain falls, it percolates through the ground, dissolving materials out of the soil. This alters its pH and introduces other materials into the water. If the soil is alkaline, the water's acidity will be neutralized, but if it is acid, the acidity of the water may increase. This water is used by plants and trees for their sustenance, and eventually flows into rivers and lakes. There have been various reports indicating that streams and lakes have been getting more acidic in recent years, although the effects seem to be highly variable and not closely correlated with releases of sulfur dioxide and nitrogen oxides. A study of Adirondacks lakes between 1975 and 1985 found that one-quarter became more acid, one-quarter (including some less than 5 miles away) became less acid, and the rest were unchanged. These findings were clearly inconclusive.

Among other things, acid rain makes lakes unlivable for fish and other aquatic life and destroys forests. On both of these matters, the evidence is highly complex and took many years to develop, and it is still not completely unequivocal. But by now most scientists are convinced that the effects are real and serious in some areas.

One of the problems is that it is difficult to be certain about any specific area. Of the 50,000 lakes in the northeastern United States, 220, mostly in the Adirondacks mountains of New York, have no fish because the water is too acidic. Since tourism is very important in the Adirondacks region, this has had serious economic consequences that may be ascribed to acid rain, and the residents are very upset about it. On the other hand, there is no evidence that there ever were fish in those lakes. The surrounding soil is naturally highly acidic, which is surely at least partially responsible for the lakes' acidity. Moreover, there is no indication that the acidity has been changing in recent years.

A National Academy of Sciences committee[3] investigated the problem by estimating the acidity prior to 1800. They analyzed fossil remains of microorganisms in the lake bottom sediments. Of the nine Adirondacks lakes for which these data and information on fish populations are available, six showed clear signs of increased acidity. For example, the pH of Big Moose Lake fell from 5.8 prior to 1800 to 4.9 at present, and the number of fish species in the lake declined from 10 in 1948 to 5 or 6 in 1962. It would normally take many hundreds or thousands of years for this large a change to take place through natural processes. The National Academy of Sciences Committee concluded that acid rain is responsible for these changes.

The problem of forest destruction is at least equally complicated. The most elaborate study was done in Germany, where a serious blight on trees has been in action.[4] In 1982 it affected 8% of all West German trees, and by 1987, 52% were affected. Since Germans revere their forests, this is a high-priority issue for them. At the time when the political issue of installation of Pershing nuclear missiles on German soil reached its highest pitch, a poll found that the death of forests was the greatest concern of the German public.

A study of one particular forest concluded that the trees were under stress from a variety of factors but acid rain was "the straw that broke the camel's back." It dissolved aluminum out of the soil—about 5% of all soil is aluminum—and this toxic element was picked up by the roots of trees. Not only did the toxicity of aluminum cause direct damage to the trees, but aluminum was picked up instead of calcium and magnesium, which are crucial to a tree's nutrition. The problem was compounded by the nitric acid in the rain acting as fertilizer to accelerate the trees' growth at a time when important nutrients were lacking. Because of this stress, the trees were succumbing to what would normally be nonlethal attacks by insects compounded by drought.

Trees are suffering blight in many parts of the world, and acid rain is suspected of contributing to the problem. A prime example are spruce trees in the Appalachian mountains, which are dying off from New England to

North Carolina. In this situation, the trees are at a high altitude where they are engulfed in a mist a large fraction of the time. Acidity in the mist is believed to be an important contributor to the damage—acidity that has the same origin as acid rain, mainly emissions from coal-burning plants.

Some of the most important problems caused by acid rain are political. The sulfur dioxide and nitrogen oxides that cause acid rain originate far away, in other states or in other countries. The acid rain that is damaging lakes and forests in the eastern United States and Canada originates in coal-burning power plants in the Midwest. (Midwestern lakes and forests are not damaged by these emissions because soil in that region is naturally less acidic.) The acid rain that damages lakes and forests in Scandinavia originates in Britain and Western Europe, and the latter also contributes to damaging German forests. It is only natural for people to be very upset by losses caused by others, and they are demanding action.

Acid rain from U.S. sources is a top priority political issue in Canada. In any meetings between leaders of the two countries, the Canadians insist that it be high on the agenda. They insist on U.S. action to reduce sulfur dioxide emissions, and because of the high value we place on Canadian friendship, they will probably get it. In 1989, the Bush Administration introduced new clean air legislation that requires cutting sulfur dioxide emissions in half by the end of the century. It is estimated that this will increase the cost of electricity from coal burning by about 20%. This still does not face the problem of nitrogen oxides, for which there is no very effective control technology.

AIR POLLUTION

The greenhouse effect and acid rain have received more media attention and hence more public concern than general air pollution. This is difficult to understand, because the greenhouse effect causes only economic disruption, and acid rain kills only fish and trees, whereas air pollution kills people and causes human suffering through illness.

We have already described the processes that produce sulfur dioxide and nitrogen oxides, which are important components of air pollution as well as the cause of acid rain. But many other processes are also involved in burning fossil fuels. When carbon combines with oxygen, sometimes carbon monoxide, a dangerous gas, is produced instead of carbon dioxide. Thousands of other compounds of carbon, hydrogen, and oxygen, classified as hydrocarbons or volatile organic compounds, are also produced in the burning of fossil fuels. During combustion, some of the carbon remains unburned, and some other materials in coal and oil are not combustible; these come off as very small solid particles, called particulates, which are typically less than one ten thousandth of an inch in diameter, and float around in the air for many days. Smoke is a common term used for particulates large enough to be visible. Some of the organic compounds formed in the combustion process attach to these particulates, including some that are known to cause cancer. Coal contains trace amounts of nearly every element, including toxic metals like beryllium, arsenic, cadmium, selenium, and lead, and these are released in various forms as the coal burns.

All of the above pollutants are formed and released directly in the combustion process. Some time after their release, nitrogen oxides may combine with hydrocarbons in the presence of sunlight to form ozone, one of the most harmful pollutants. Or other compounds may form, such as PAN, which is best known as the cause of watering eyes in Los Angeles smog.

Let us summarize some of the known health effects of these pollutants:[5]

- Sulfur dioxide is associated with many types of respiratory diseases, including coughs and colds, asthma, bronchitis, and emphysema. Studies have found increased death rates from high sulfur dioxide levels among people with heart and lung diseases.
- Nitrogen oxides can irritate the lungs, cause bronchitis and pneumonia, and lower resistance to respiratory infections such as influenza; at higher levels it can cause pulmonary edema.
- Carbon monoxide bonds chemically to hemoglobin, the substance in the blood that carries oxygen to the cells, and thus reduces the amount of oxygen available to the body tissues. Carbon monoxide also weakens heart contractions, which further reduces oxygen supplies and can be fatal to people with heart disease. Even at low concentrations it can affect mental functioning, visual acuity, and alertness.
- Particulates, when inhaled, can scratch or otherwise damage the respiratory system, causing acute and/or chronic respiratory illnesses. Depending on their chemical composition, they can contribute to other adverse health effects. For example, benzo-a-pyrene, well recognized as a cancer-causing agent from its effects in cigarette smoking, sticks to surfaces of particulates and enters the body when they are inhaled.
- Hydrocarbons cause smog and are important in the formation of ozone.
- Ozone irritates the eyes and the mucous membranes of the respiratory

tract. It affects lung function, reduces ability to exercise, causes chest pains, coughing, and pulmonary congestion, and damages the immune system.

- Volatile organic compounds include many substances that are known or suspected to cause cancer. Prominent among these is a group called *polycyclic aromatic*, which includes benzo-a-pyrene mentioned above.
- Toxic metals have a variety of harmful effects. Cadmium, arsenic, nickel, chromium, and beryllium can cause cancer, and each of these has additional harmful effects of its own. Lead causes neurological disorders such as seizures, mental retardation, and behavioral disorders, and it also contributes to high blood pressure and heart disease. Selenium and tellurium affect the respiratory system, causing death at higher concentrations.

It is well recognized that toxic substances acting in combination can have much more serious effects than each acting separately, but little is known in detail about this matter. Information on the quantities of air pollutants required to cause various effects is also very limited. However, there can be little doubt that air pollution is a killer.

The clearest evidence linking air pollution to increased mortality comes from several catastrophic episodes[6] in which a large number of excess deaths occurred during times of high pollution levels, in all cases caused by coal burning in association with unfortunate weather conditions. In a December 1930 episode in the Meuse Valley of Belgium, there were 60 excess deaths and 6,000 illnesses. In an October 1948 episode in Donora, Pennsylvania, there were 20 deaths (versus 2 normally expected) in a 4-day period during which 6,000 of the 14,000 people in the valley became ill. There were at least eight episodes in London between 1948 and 1962 in each of which hundreds of excess deaths were recorded, the largest in December 1952 when 3,500 died. There were three episodes in New York City involving over a hundred deaths, one in November 1953 causing 360 deaths, another in January-February 1963 leading to 500 deaths, and a third in November 1966 responsible for 160 deaths. In all of these cases, the mortality rates rose sharply when measured air pollution levels reached very high values, and fell when the latter declined.

The best method for establishing a connection between "normal" levels of air pollution and premature mortality is through comparison of mortality rates between different geographic areas with different average air pollution levels. Of course there are other factors affecting mortality rates that vary with geographic area, like socioeconomic conditions; the data must be analyzed thoroughly to eliminate these factors. As an example, Fig. 1 shows a plot of annual mortality rates for males aged 50-69 in various census tracts of Buffalo, New York, during 1959-1961, versus average annual income and average air pollution.[7] It is evident from Fig. 1 that for each income range, mortality rates increased with increasing pollution level. A mathematical analysis separates the two effects, giving the risk of air pollution alone. There have been a number of other such studies,[8] comparing various cities in the United States, all the counties in the United States, various cities in England, and so on.

In addition, there have been a number of studies[9] of mortality rates in a given city, especially New York and London (also in several other American cities and Tokyo), on a day-by-day basis, correlating them with air pollution levels. In these there are no complications from socioeconomic factors, since these do not vary on a day-to-day basis. However, there are weather factors that must be removed by mathematical analysis in order to determine the effects of air pollution alone.

These studies have established strong correlations in timing between elevated air pollution levels and mortality rates. There are also numerous studies[10] of temporary illness, involving hospital admissions, questionnaires, measurements of pulmonary function, and so on, in New York,

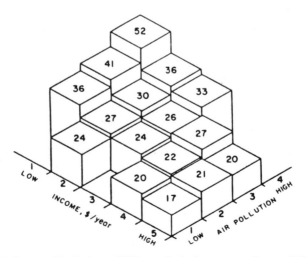

Fig. 1. Mortality rate (deaths/year/1000 population) among males aged 50-69 versus annual income and air pollution in the section of Buffalo, New York, in which they live.

London, Chicago, five Japanese cities, Rotterdam, Oslo, and others, all indicating strong correlations with abnormally high air pollution levels.

The U.S. Department of Energy's Office of Health and Environmental Research sponsored a multiyear study by a Harvard University research group to evaluate all of the available studies. Its conclusion was that air pollution is probably causing about 100,000 deaths per year in the United States.[11] These deaths are principally from heart and lung disease. In addition it is estimated that air pollution causes about 1,000 cancer deaths per year.[6]

The estimate of 100,000 deaths per year means that one American out of 30 dies as a result of air pollution. Most environmental agents that get abundant media attention and public concern, such as Alar in apples, pesticides that have been banned, PCBs, and formaldehyde, give those exposed less than one chance in 100,000 of dying from their effects. We see that air pollution, which gives 1 chance in 30, is thousands of times more harmful.

While the evidence for health effects from air pollution is undeniable, reaching an understanding of them has proved to be a very difficult task. Historically, the pollutants most easily and therefore most frequently measured were sulfur dioxide and suspended particulate matter; thus, nearly all correlation studies were based on them. Until the late 1970s, it was widely assumed that these were the materials actually responsible for the health damage. Yet animals exposed to very high levels of these materials for long time periods showed no ill effects. Moreover, men occupationally exposed to 100 times the normal outdoor levels of sulfur dioxide from refrigeration sources, oil refineries, and pulp and paper mills were not seriously affected. Occupational settings where suspended particulate levels are 100 times higher than the outdoor average revealed no important health effects.[10] In response to these findings, in the mid-1970s there was a strong trend toward designating sulfate particulates, resulting from chemical reactions of sulfur dioxide with other chemicals in air, as the culprit, and some still continue to support that viewpoint; but animals show no ill effects even from prolonged exposure to relatively high levels of sulfate particulates.

One could concentrate on other components of air pollution as possible sources of the health effects, and there are plenty of candidates to choose from. Most of them, including the great majority of volatile organic compounds, have not been investigated as causes of health effects. But there is no evidence or strong body of opinion that any one substance is the major culprit in air pollution. Health effects probably arise from complicated interactions of many pollutants acting together. We may never understand the process in any detail.

The greatest difficulty in trying to tie down causes is that air pollution doesn't kill healthy people in one fell swoop. It rather continuously weakens the respiratory and cardiovascular systems over many decades until they collapse under one added insult. This explains why there are no mortality effects on the occupationally exposed, on animals, and on college students used as volunteer subjects in controlled tests.[10]

Since we do not know what components of air pollution cause the health effects, it is impossible to know what pollution control technologies will be effective in averting them. The often-heard statement "we can clean up coal burning" involves a large measure of wishful thinking.

How much of this air pollution can be averted by use of nuclear power? None of the pollutants discussed above are released by nuclear reactors. Coal burning, which now generates most of our electricity, is by far the most polluting process. According to EPA estimates,[5] fossil fuel burning in electric power plants produces 64% of all U.S. releases of sulfur dioxide, 27% of the particulates, and 31% of the nitrogen oxides, but less than 1% of the carbon monoxide or hydrocarbons. It is therefore reasonable to estimate that this is causing 30,000 of the 100,000 deaths per year caused by air pollution. Industrial fuel combustion, which is being rapidly replaced by electricity, produces 12% of the sulfur dioxide, 10% of the particulates, 22% of the nitrogen oxides, and 5% of the volatile organic compounds. Averting these emissions would make a very substantial contribution to the solution of our air pollution problems. If electric cars were to be successful and more electrically powered buses and railroads were used, further great gains would be achieved since transportation is responsible for 84% of the carbon monoxide, 41% of the volatile organic compounds, and 40% of the nitrogen oxides.

We are spending an estimated $30 billion per year to reduce air pollution, and a substantial fraction of it is due to coal-burning power plants. Replacing them with nuclear plants would therefore save many billions of dollars per year in this area alone.

Other Environmental Effects of Fossil Fuels[12]

Sixty percent of our coal is now strip-mined. That is, huge earth-moving machines strip off the covering soil to reach the coal and then scoop it up and load it into trucks. In a single bucket load these machines pick up 300 tons. They sometimes remove close to 200 feet of covering soil to reach the coal. This is earth moving on a gigantic scale, and in the process, the land is badly scarred. Most states now have laws requiring that the land contour be restored and revegetated. These laws have done a lot to improve the situation, but the restoration often leaves much to be desired. Our largest coal reserves are in the Wyoming-Montana region, where restoration is particularly difficult because of sparse rainfall. Strip-mine operators are required to post a bond as evidence of intent to restore the land after mining, but it is said that no coal company has ever gotten its reclamation performance bond returned in Montana.[12] Over a million acres of strip-mined land is awaiting reclamation, and new land is being strip-mined at a rate of 65,000 acres per year.

The remaining 40% of our coal comes from underground mines, and that percentage will eventually have to increase as locations for strip mining run out—the great majority of our coal reserves require underground mining. One of the environmental impacts of this endeavor is acid drainage from abandoned mines. Water seeping in reacts with sulfur compounds to produce sulfuric acid, which eventually seeps out and gets into streams, making them acidic. This kills fish and makes the water unfit for drinking, swimming, and many industrial applications. There are methods for preventing this acid mine drainage, but they are quite expensive and are not generally being implemented.

Another environmental impact of underground mining is land subsidence (ground on the surface moving downward as the abandoned mines below cave in), causing buildings on the surface to crack or even to be completely destroyed. Something like one-fourth of the 8 million acres that are above coal mines have subsided. About 7% of this subsidence has been in cities, where damage is very high and tragic to the homeowners. Subsidence in rural areas can change drainage patterns and make land unfit for farming. At the very least it scars the land. Laws governing mineral rights relieve mining companies of any responsibility for damage done by subsidence.

Another environmental impact of underground mining is fires that start by accident and are very difficult to put out. Some have smoldered for many decades. They release smoke laden with air pollutants, and their heat kills the vegetation. Of course, they also destroy a lot of coal. In 1983 there were 261 uncontrolled mine fires smoldering in the United States.

Coal is often washed just outside the mine to remove foreign materials, and the waste material from this washing is piled up in unsightly heaps. In 1983 there were 177,000 acres of these waste banks in the United States, the great majority in Appalachia. Many of these waste banks catch fire and burn, serving as a source of air pollution.

Mining coal is among the most unpleasant of occupations. A miner is in constant intimate contact with dirt, often without room to stand up, and engulfed in dust. Idealists have called it a job unfit for humans. Although there have been many improvements in recent years, it is still one of the most dangerous occupations, regularly killing over 100 men per year in the United States. (Early in this century it was killing well over 1,000 per year.)

But the most important health impact of coal mining is black lung disease, a name derived from the fact that, on autopsy, the lungs of coal miners are invariably found to be black. That disease, which causes lots of discomfort, is not fatal. However, it leaves miners exceptionally susceptible to emphysema and a variety of other lung diseases.[13] Underground miners have more than 20 times the normal risk of death from silicosis or pneumoconiosis, lung diseases caused by exposure to dust, and about 2.5 times the normal risk of dying from bronchitis, pneumonia, or tuberculosis. At younger ages, coal miners are healthier than the average person their age, but as they age the situation reverses; coal miners die an average of 3 years younger than the rest of the population of the same socioeconomic status.

The most publicized environmental effect of using petroleum as a fuel is oil spills, highlighted by the spill of 40,000 tons of oil from the tanker *Exxon Valdez* off the coast of Alaska in 1989. Although over a billion dollars was spent in the clean-up, many of the beaches were ruined and numerous species of aquatic animals suffered damage that will not be healed for decades. But by world standards this was not a large oil spill. In 1979, the *Atlantic Empress* was involved in a collision off the coast of Tobago in the Caribbean, spilling 305,000 tons, and in 1978 the *Amoco Cadiz* ruined many miles of French beaches with a spill of 237,000 tons. There are lots of smaller spills. U.S. tankers alone spilled an average of 215,000 tons per year in 1970-1974, and 380,000 tons per year in 1975-1979. At any given time, over 100 million tons of oil is being transported by ships, so it is not surprising that some of it occasionally ends up in the water.

Accidents on land can also spill oil into the oceans. The most spectacular case of this type was in Campeche Bay, Mexico, in 1979, where a well could not be capped for 280 days, during which it spilled 700,000 tons of oil into the Gulf of Mexico, doing heavy damage to aquatic life.

Our news media give much more publicity to spills off U.S. shores. Perhaps the most publicized was the 1969 spill off the California coast near Santa Barbara, in which 7,000 tons went into the water before the well was capped. The damage the oil did to the beautiful beaches there stopped offshore drilling in that region for 20 years.

It seems clear that as long as we use oil heavily, we will pollute the oceans with it, and damage the local aquatic life. With the biggest spills to date off our shores being 7,000 tons in 1969 and 40,000 tons in 1989, we have perhaps been abnormally lucky.

CONCLUSIONS

In the last chapter, we showed that lots of new power plants will be needed in the United States in the near future, and that they will have to be nuclear or burners of fossil fuels. In this chapter we have reviewed some of the drawbacks to the latter, and we have seen that they are very substantial. These should be kept in mind as we consider the environmental problems with nuclear power in later chapters.

Chapter 4 / IS THE PUBLIC READY FOR MORE NUCLEAR POWER?

We have seen that we will need more power plants in the near future, and that fueling them with coal, oil, or gas leads to many serious health, environmental, economic, and political problems. From the technological point of view, the obvious way to avoid these problems is to use nuclear fuels. They cause no greenhouse effect, no acid rain, no pollution of the air with sulfur dioxide, nitrogen oxides, or other dangerous chemicals, no oil spills, no strain on our economy from excessive imports, no dependence on unreliable foreign sources, no risk of military ventures. Nuclear power almost completely avoids all the problems associated with fossil fuels. It does have other impacts on our health and environment, which we will discuss in later chapters, but you will see that they are relatively minor.

However, regardless of such rational arguments, we must recognize that acceptance of nuclear power is largely an emotional issue. Nuclear power cannot have a future in the United States unless the public at large is ready to accept it. Pollsters have been busily trying to determine the status of this question. Let's examine what they have found.

WHAT THE POLLS SAY

A Gallup poll taken in February 1989 asked "How important do you think nuclear energy plants will be in providing this nation's electricity needs in the years ahead?" The same question was then asked about coal-burning plants. The results are given below for the first question, with the results for coal-burning plants in parentheses.

a.	Very important	45%	(29%)
b.	Somewhat important	34%	(37%)
c.	Not too important	10%	(17%)
d.	Not at all important	6%	(8%)
e.	Don't know	5%	(9%)

Combining a with b, we see that 79% thought nuclear would be important, while only 66% thought coal, which now provides most of our electric power, would be important. Combining c with d, only 16% thought nuclear would be unimportant versus 25% for coal. Moreover, younger people are more favorably disposed toward nuclear, with 81% in the age range 18-34 versus 75% over age 55 believing that nuclear would be important.

In July 1989, the polling organization TeleNation Market Facts asked a cross section of the American public "How important a role *should* nuclear energy play in the U. S. Department of Energy's National Energy Strategy for the future?" Responses were

Very important	50%
Somewhat important	31%
Not too important	8%
Not at all important	8%
Don't know	3%

The results of a May 1989 survey by another professional polling organization, Cambridge Reports, were not quite as impressive, but still encouraging. It asked a representative sample of the American public "How important a role *should* nuclear energy play in meeting America's future energy needs?" The results were

Very important	38%
Somewhat important	31%
Not too important	13%
Not at all important	14%
Don't know	4%

At the same time, Cambridge Reports asked another representative sample "How important do you think nuclear energy plants *will be* in meeting this nation's energy needs in the years ahead?" Replies were

Very important	50%
Somewhat important	27%
Not too important	10%
Not at all important	9%
Don't know	4%

To summarize these surveys, 69% to 81% of the public think nuclear power *should be* important, while 16% to 27% think it *should be* unimportant. When asked whether it *will be* important, 77% to 79% said yes, while 16% to 19% said no.

These surveys refer somewhat indefinitely to the future. But what about the very near future that must be actively planned for now? In May 1989, Cambridge Reports asked "Which one energy source do you think will be our primary source of electricity 10 years from now?" The interviewer did not name any sources. The results were

Nuclear energy	28%
Solar energy	18%
Hydroelectric	6%
Coal	6%
Oil	6%
Natural gas	4%
Fusion	3%
Wind	1%
Other	5%
Don't know	23%

Note that nuclear energy is by far the leader, and its lead over coal, oil, and natural gas, which experts would all agree are the only practical alternatives for the near future, is especially impressive.

In November 1989, Cambridge Reports asked "Do you think the nation's need for nuclear energy as part of the total energy mix will increase in the years ahead?" Of those polled, 77% said yes, 15% said no, and 8% were not sure.

One might still wonder whether the public is in favor of constructing *any* new power plants. The May 1989 Cambridge Reports poll also asked how serious a problem are energy supplies in the United States. Responses were

Very serious	48%
Somewhat serious	32%
Not very serious	11%
Not serious at all	6%
Don't know	3%

The 48% vote for "very serious" is up from about 33% during the 1983-1986 time period.

All of this leads us to believe that the overwhelming majority of the American public will not be surprised or offended by more nuclear power plants being built in the near future and is ready to accept them. Since there has been little publicity about the new developments, these attitudes are based on current nuclear power plants. When the public is informed about the new super-super safe plants to be described in Chapter 10, it should be even more favorably inclined.

While public support of nuclear power has only recently been turning favorable, the scientific community has always been steadfastly supportive. In 1980, at the peak of public rejection, Stanley Rothman and Robert Lichter, social scientists from Smith College and Columbia University, respectively, conducted a poll of a random sample of scientists listed in *American Men and Women of Science*, The "Who's Who" of scientists.[1] They received a total of 741 replies. They categorized 249 of these respondents as "energy experts" based on their specializing in energy-related fields rather broadly defined to include such disciplines as atmospheric chemistry, solar energy, conservation, and ecology. They also categorized 72 as nuclear scientists based on fields of specialization ranging from radiation genetics to reactor physics. Some of their results are listed in Table 1.

From Table 1 we see that 89% of all scientists, 95% of scientists involved in energy-related fields, and 100% of radiation and nuclear scientists favored proceeding with the development of nuclear power. Incidentally, there were no significant differences between responses from those em-

TABLE 1
HOW SHOULD WE PROCEED WITH NUCLEAR POWER DEVELOPMENT?

	All scientists %	Energy experts %	Nuclear experts %
Proceed rapidly	53	70	92
Proceed slowly	36	25	8
Halt development	7	4	0
Dismantle plants	3	1	0

ployed by industry, government, and universities. There was also no difference between those who had and had not received financial support from industry or the government.

Another interesting question was whether the scientists would be willing to locate nuclear plants in cities in which they live (actually, no nuclear plants are built within 20 miles of heavily populated areas). The percentage saying that they were willing was 69% for all scientists, 80% for those in energy-related sciences, and 98% for radiation and nuclear scientists. This was in direct contrast to the 56% of the general public that said it was *not* willing.

Rothman and Lichter also surveyed opinions of various categories of media journalists and developed ratings for their support of nuclear energy. Their results are shown in Table 2.

We see that scientists are much more supportive of nuclear power than journalists, and press journalists are much more supportive than the TV people who have had most of the influence on the public, even though they normally have less time to investigate in depth. There is also a tendency for science journalists to be more supportive than other journalists.

In summary, these Rothman-Lichter surveys show that scientists have been much more supportive of nuclear power than the public or the TV reporters, producers, and journalists who "educate" them. Among scientists, the closer their specialty to nuclear science, the more supportive they are. This is not much influenced by job security considerations, since the level of support is the same for those employed by universities, where tenure rules protect jobs, as it is for those employed in industry. Moreover, job security for energy scientists is not affected by the status of the nuclear industry because they are largely employed in enterprises competing with nuclear energy. In fact, most nuclear scientists work in research on radiation and the ultimate nature of matter, and are thus not affected by the status of

TABLE 2
SUPPORT FOR NUCLEAR ENERGY[a]

Category	Number surveyed	Support rating
Nuclear scientists	72	7.9
Energy scientists	279	5.1
All scientists	741	3.3
Science journalists	42	1.3
Prestige press journalists	150	1.2
Science journalists at *New York Times, Washington Post,* TV networks	15	0.5
TV reporters, producers	18	−1.9
TV journalists	24	−3.3

[a]Scale runs from +10 for perfect to −10 for complete rejection.

the nuclear power industry. Even among journalists, those who are most knowledgeable are the most supportive. The pattern is very clear—the more one knows about nuclear power, the more supportive one becomes.

But the attitude of scientists is largely irrelevant. The decision on acceptability of nuclear power will be made by the public. The most important point is that the public is now becoming supportive. If that support is maintained and continues to grow, the future of nuclear power is certain to be bright.

Strong public support for nuclear power is not a new phenomenon in the United States. In the 1960s, the public viewed nuclear energy as the great new wave of the future, the answer to all energy supply problems. Candidate communities vied for the honor of being chosen as a location for a nuclear plant. Mayors and governors offered tax concessions and pulled political strings to get them.

All of that changed when groups opposed to nuclear power formed and gained support from the media in depicting it as a dangerous technology operated by incompetents. Public opinion was turned around, with disastrous consequences. It has taken many years for the nuclear industry to recover public support.

This support could be lost again if the opposing groups and the media were to perform as they did in the 1970s and early 1980s. But there is evidence that this may not happen. Environmental groups are now truly concerned about the greenhouse effect, acid rain, and air pollution and, on weighing the alternatives, are becoming more opposed to coal burning than to nuclear energy.

According to polls by Cambridge Reports, in the 1983-1986 time period, which was 4-7 years after the Three Mile Island accident but before Chernobyl, 60% of the public recalled seeing or reading news stories about nuclear energy within the previous 6 months, and over two-thirds of these stories were "mostly unfavorable." However, in May 1989, only 51% recalled stories and only 25% viewed them as unfavorable.

In the 20 polls taken between 1983 and 1989, this May 1989 poll was the first in which less than half of the stories were unfavorable. This gives grounds for optimism. Since many stories are neutral, there are still more unfavorable than favorable stories, so the decrease in total number of stories is also a positive development. Perhaps the media and the public are ready to base decision making on nuclear power on scientific information. If so, decisions will be made on the basis of material we will explore in the remainder of this book.

But public support of nuclear power is fragile, and suspicion abounds. In a February 1989 Gallup poll, people were asked whether selecting nuclear power for large-scale use was a good choice, a realistic choice, or a bad choice. Replies were

Good choice	19%
Realistic choice	50%
Bad choice	25%

It seems like the public is not in love with nuclear power but is ready to accept it as the least of the available evils. The public prefers to avoid risks in any shape or form but is coming to recognize that some risk is unavoidable. We will explore this matter in Chapter 8.

This February 1989 Gallup poll asked people how they would react to having a nuclear power plant in their community. Replies were

Favor nuclear plant	17%
Oppose nuclear plant	23%
Reserve judgment	59%

This hesitancy to have a plant in one's own community is a pervasive attitude known as the NIMBY syndrome—Not in my back yard. It extends far beyond the nuclear industry to any industrial activity that is viewed as something less than pristine. It is therefore encouraging to see that a strong majority of the public is at least willing to reserve judgment on a nuclear power plant in their community. In 1981, 56% were opposed.

Perhaps the best way to summarize the poll results presented in this chapter is to conclude that the public is receptive to, and even supportive of, nuclear power, but it is suspicious and can easily be swayed in either direction. Presumably, the heavy majority that is reserving judgment is waiting for more information on the subject. Providing that information is my main goal in writing this book.

PUBLIC MISUNDERSTANDING

I have been doing research and teaching on the health and environmental impacts of nuclear power for the past 17 years and have been constantly dismayed by the vast gulf of misunderstanding by the public. Perhaps the most important misunderstanding is about the dangers of radiation. The public views radiation as something highly mysterious, very complex, and poorly understood. Actually, it is one of the simplest and best understood of all environmental agents, far better understood, for example, than the biological actions of sulfur dioxide, nitrogen oxides, or any of the other chemical agents discussed in Chapter 3. The next chapter represents my effort to clear up the public's misunderstanding of the hazards of radiation.

Another important misunderstanding is the danger from reactor meltdown accidents. Many people view such an accident as the ultimate disaster, picturing tens of thousands of dead bodies strewn about the landscape, something like what may be expected from a nuclear bomb attack. Actually, it is impossible for a reactor meltdown to cause anything approaching that level of disaster. In fact, deaths among the public from a meltdown accident would be similar to those from the air pollution caused by coal burning. They would be predominantly among the elderly, and only very rarely would they be recognizably connected with the accident. The major difference from the air pollution analogy is that there would be only a tiny fraction as many deaths. For the number of deaths from reactor meltdown accidents to be equal to the number caused by coal-burning air pollution, there would have to be a complete meltdown accident somewhere in the United States every few days! But after more than 30 years of nuclear power, we haven't even had the first such accident yet.

Chapter 6 explains reactor meltdown accidents, including their potential causes, their estimated effects, and estimates of how often they may be expected. This is followed in Chapter 7 by a description of the Chernobyl accident in the Soviet Union, including a definitive answer to the question: Can such an accident happen here?

The third major misunderstanding results from the failure of the public to quantify risks and put them into perspective with other risks. Chapter 8 represents an effort to lead the reader through this process, allowing the risks of nuclear power to be expressed in terms of extra cigarettes consumed by a regular cigarette smoker, extra weight gained by an overweight person, or driving in a small car rather than a midsize car. We then turn to the question of cost effectiveness of life-saving measures, comparing the number of deaths that are being averted by spending a given amount of money to improve the safety of nuclear reactors and of radioactive waste from the nuclear industry with the number that *could be* averted by spending that same money on medical screening or highway safety programs, or even on reducing our radiation exposure from radon in homes. The facts leading to the conclusions I draw in Chapter 8 were absolutely astounding to me when I discovered them. Perhaps they will serve as eye-openers to the reader.

The fourth major misunderstanding is of the hazards associated with radioactive waste, which is the subject of Chapters 11 and 12. Much of the public views this as an unsolved problem, with horrible consequences a distinct possibility if it is not solved satisfactorily. Actually, it is a rather trivial technological problem, and it can be shown that the health risks are trivial compared with those due to the waste from burning fossil fuels. Elucidating these matters is the principal agenda of Chapters 11 and 12, but the story involves several side issues that are also covered there.

Even if these four major areas of misunderstanding are cleared up, the public will still not accept nuclear power if the financial cost is too high. As of now, the costs are too high, or at least somewhat higher than the costs of electricity derived from burning coal. This situation has only come to pass recently—until the mid-1980s, nuclear power was cheaper. The reasons for that turnabout are the topics for Chapter 9.

The solution to the cost problem is given in Chapter 10, which describes the new generation of nuclear power plants being developed for the 1990s and beyond. They will not only provide electricity at a lower cost than coal-generated electric power, but they will also be a thousand times safer than plants of the present generation. Why this is possible and how it will be done will be described in some detail.

Some other misunderstandings of nuclear energy that have received less attention recently, the hazards of plutonium and the possible role of plutonium from the nuclear industry in making bombs for terrorists or for nations that do not now have nuclear weapons, are the topics of Chapter 13.

But even if all of the misunderstandings are cleared up, some people would still hesitate to accept nuclear power because they prefer solar energy instead. Is that a real option? Or will it become a viable option in the foreseeable future? Is it really as environmentally benign as most people believe it to be? All of these matters are considered in Chapter 14.

Unfortunately, some of the discussions must become rather technical. I have done my best to avoid this without leaving out essential points. Some of the technical details have been relegated to the Appendixes. But, to a large extent, each chapter stands by itself, and in most cases, can be read without having read the previous chapters.

With these preliminaries out of the way, we are ready to begin our discussion of the public's misunderstandings about nuclear power. We begin with the question, "How dangerous is radiation?"

Chapter 5 / HOW DANGEROUS IS RADIATION?

The most important breakdown in the public's understanding of nuclear power is in its concept of the dangers of radiation. What is radiation, and how dangerous is it?

Radiation consists of several types of subatomic particles, principally those called *gamma rays, neutrons, electrons,* and *alpha particles,* that shoot through space at very high speeds, something like 100,000 miles per second. They can easily penetrate deep inside the human body, damaging some of the biological cells of which the body is composed. This damage can cause a fatal cancer to develop, or if it occurs in reproductive cells, it can cause genetic defects in later generations of offspring. When explained in this way, the dangers of radiation seem to be very grave, and for a person to be struck by a particle of radiation appears to be an extremely serious event. So it would also seem from the following description in what has perhaps been the most influential book from the opponents of nuclear energy[1]:

When one of these particles or rays goes crashing through some material, it collides violently with atoms or molecules along the way. . . . In the delicately balanced economy of the cell, this sudden disruption can be disastrous. The individual cell may die; it may recover. But if it does recover, after the passage of weeks, months or years, it may begin to proliferate wildly in the uncontrolled growth we call cancer.

But before we shed too many tears for the poor fellow who was struck by one of these particles of radiation, it should be pointed out that every person in the world is struck by about 15,000 of these particles of radiation every second of his or her life,[2] and this is true for every person who has ever lived and for every person who ever will live. These particles, totalling 500 billion per year, or 40 trillion in a lifetime, are from natural sources. In addition, our technology has introduced new sources of radiation like medical X-rays—a typical X-ray bombards us with over a trillion particles of radiation.

With all of this radiation exposure, how come we're not all dying of cancer? The answer to that question is *not* that it takes a very large number of these particles to cause a cancer. As far as we know, every single one of them has that potential; as we are frequently told, "no level of radiation is perfectly safe." What saves us, rather, is that the probability for one of these particles to cause cancer is very low, about 1 chance in 30 quadrillion (30 million billion, or 30,000,000,000,000,000)! Every time a particle of radiation strikes us, we engage in a fatal game of chance at those odds. However, this is not unique to radiation; we are engaged in innumerable similar games of chance involving chemical, physical, and biological processes that may lead to any form of human malady, and the one involving radiation has odds much more favorable to us than most. Only about 1% of fatal human cancers are caused by the 30 trillion particles of radiation that hit us over a lifetime (this estimate does not include the effects of radon, to be discussed below), while the other 99% are from losing in one of these other games of chance.

Of course every extra particle that strikes us increases our cancer risk, so many people feel that they should go to great lengths to avoid extra radiation. If that is your attitude, there are many things you can do. You can reduce it 10% by living in a wood house rather than a brick or stone house,[3] because brick and stone contain more radioactive materials like uranium, thorium, and potassium. You could reduce it 20% by building a thick lead shield around your bed to reduce the number of hits while you sleep, or you could cut it in half by wearing clothing lined with lead like the cover dentists drape over you when they take X-rays.

But most people don't bother with these things. Rather, they recognize that life is full of risks. Every time you take a bite of food, it may have a chemical that will initiate a cancer, but still people go on eating, more than necessary in most cases. Every ride or walk we take could end in a fatal accident, but that doesn't keep us from riding or walking. Similarly, the sensible attitude most of us take is not to worry about a little extra radiation; after all, 1 chance in 30 quadrillion is pretty good odds!

The moral of this story is that hazards of radiation must be treated quantitatively. If we stick to qualitative reasoning alone, we can easily conclude that nuclear power is bad—it leads to radiation exposure which can cause cancer. The trouble with this is that, by a similar type of qualitative reasoning, just about anything else we do can be shown to be harmful: coal or oil burning causes air pollution which kills people, so coal or oil burning is bad; using natural gas leads to explosions which kill people, so burning gas is bad; and so on. Any discussion of dangers from radiation must include numbers; otherwise, it can be as completely deceptive as the quote above about the tragedy of being struck by a single particle of radiation. But how often do stories we hear about radiation include numbers?

MEET THE MILLIREM

In order to discuss radiation exposure quantitatively, we must introduce the unit in which it is measured, called the *millirem,* abbreviated mrem. One millirem of exposure corresponds to being struck by approximately 7 billion particles of radiation, but it takes into account variations in health risks with particle type and size of person. For example, a large adult and a small child standing side by side in a field of radiation would suffer roughly the same cancer risk and hence would receive the same dose in millirems, although the adult would be struck by many more particles of radiation being a larger target. In nearly all of our discussions about radiation, we will be considering doses below about 10,000 mrem, which is commonly referred to as low-level radiation.

We frequently hear stories about incidents in which the public is exposed to radiation; radioactive material falling off a truck; contaminated water leaking out of a tank or seeping out of a waste burial ground; a radioactive source used for materials inspection being temporarily misplaced; malfunctions in nuclear plants leading to releases of radioactivity; and so on. Perhaps a hundred of these stories over the past 45 years have received national television coverage. The thing I always look for in these stories is the radiation exposure in millirems, but it is hardly ever given.

Eventually it appears in a technical journal, or I trace it down by calls to health officials. On a very few occasions it has been as high as 5-10 mrem, but in the great majority of cases it has been less than 1 mrem. In the Three Mile Island accident, average exposures in the surrounding area[4] were 1.2 mrem—this drew the one-word banner headline "RADIATION" in a Boston newspaper. In the supposed leaks of radioactivity from a low-level waste burial ground near Moorhead, Kentucky, there were no exposures[5] as high as 0.1 mrem; yet this was the subject of a three-part series in a Philadelphia newspaper[6] bearing headlines "It's Spilling All Over the U.S.," "Nuclear Grave is Haunting KY," and "There's No Place to Hide." In the highly publicized leak from a nuclear power plant near Rochester, New York, in 1982, no member of the public was exposed to as much as 0.3 mrem.[7] Yet this was the top news story on TV network evening news for two days.

For purposes of discussion, let us say that a typical exposure in these highly publicized incidents was 1 mrem. Did these incidents really merit all this publicity? How dangerous is 1 mrem of radiation?

Perhaps the best way to understand this is to compare it with natural radiation[3]—the 15,000 particles from natural sources that strike us each second throughout life. We are constantly bombarded from above by cosmic rays showering down on us from outer space, hitting us with 30 mrem per year; from below by radioactive materials like uranium, potassium, and thorium in the ground—20 mrem/year; from all sides by radiation from the walls of our buildings (brick, stone, and plaster are derived from the ground)—10 mrem/year; and from within, due to the radioactivity in our bodies (mostly potassium)—25 mrem/year. All of these combined give us a total average dose of about 85 mrem per year from natural sources, or 1 mrem every 4 days. Thus, radiation exposures in the above mentioned highly publicized incidents are no more than what the average person receives every few days from these natural sources.

But, you might say, the extra radiation is what we should worry about because there is nothing we can do about natural radiation. Not true. The numbers given above are national averages, but there are wide variations. In Colorado and other Rocky Mountain states (Wyoming, New Mexico, Utah), where the uranium content in soil is abnormally high, and where the high altitude reduces the amount of air above that shields people from cosmic rays, natural radiation is nearly twice the national average; but in Florida, where the altitude is minimal and the soil is deficient in radioactive materials, natural radiation is 15% below the national average. Thus, the radiation exposures in the highly publicized incidents are about equal to the extra radiation you get from spending five days in Colorado. Of course, millions of people spend their whole lives in Colorado, and as it turns out, the cancer rate[8] in that state is 35% below the national average. Leukemia, probably the most radiation-specific type of cancer, occurs at only 86% of the national average rate in Colorado, and at 61% of that rate in the other high-radiation Rocky Mountain states. This is a clear demonstration that radiation is not one of the important causes of cancer. (Recall our estimate that, excluding radon, it is responsible for about 1% of all fatal cancers.)

Diagnostic X-rays are our second largest source of whole body exposure. A dental X-ray gives us about 1 mrem, and a chest X-ray gives us about 6 mrem, but nearly all other X-rays give far higher exposures[9]: pelvis, 90 mrem; abdomen, 150 mrem; spine, 400 mrem; barium enema, 800 mrem. Often a series of X-rays is taken, giving total exposures of several thousand millirems. The average American gets about 80 mrem per year[3] from this source, 80 times the exposure in the highly publicized radiation incidents. Again this diagnostic X-ray exposure is not unavoidable—much could be done to reduce it substantially without compromising medical effectiveness, and large numbers of X-rays are taken only to protect doctors and hospitals against liability suits.

There are several trivial sources of whole body radiation that give us[10] about 1 mrem: an average year of TV viewing, from the X-rays emitted by television picture tubes; a year of wearing a luminous dial watch, since the luminosity comes from radioactive materials; and a coast-to-coast airline flight, because the high altitude increases exposure to cosmic rays. Each of these activities involves about the same radiation exposure as the highly publicized incidents.

All of the above-listed sources bombard all organs of our body, but the most important source of our exposure to radiation is radon gas in our homes, which only irradiates our lungs, causing lung cancer. It gives us a cancer risk equal to that of exposing all of our body organs to 200 mrem per year. In some states, like Colorado and Iowa, the average level is 3 times this average, about 600 mrem per year. Several other areas, like an extensive one covering Altoona, Harrisburg, Lancaster, York, Allentown, Bethlehem, and Easton, Pennsylvania, and the regions around Columbus, Ohio, Nashville, Tennessee, and Spokane, Washington, have equally high radon levels. Radiation exposures in the highly publicized incidents are thus considerably lower than those received by people in these areas every day. Note that the Pennsylvania area includes the region around the Three Mile Island plant; people living near that plant get more radiation exposure from radon in their homes every day than they got from the 1979 accident. Within any area, there is a wide variation in radon levels from house to house. About 5% of us, 12 million Americans, get more than 1,000 mrem per year, and perhaps 2 million Americans get over 2,000 mrem per year from radon. In a few houses, exposures have been found to be as high as 500,000 mrem per year.

How dangerous is 1 mrem of radiation? The answer can be given in quantitative terms, with some qualifications to be discussed later, but in most situations, for each millirem of radiation we receive, our risk of dying from cancer is increased by about 1 chance in 4 million. This is the result arrived at independently by the U.S. National Academy of Sciences Committee on Biological Effects of Ionizing Radiation[11] and the United Nations Scientific Committee on Effects of Atomic Radiation.[12] The International Commission on Radiological Protection has always accepted estimates by these prestigious groups, as has the U.S. National Council on Radiation Protection and Measurements, the British National Radiological Protection Board, and similar groups charged with radiation protection in all technologically advanced nations.

This risk corresponds to a reduction in our life expectancy by 2 minutes. A similar reduction in our life expectancy is caused by[13]

- crossing streets 5 times (based on the average probability of being killed while crossing a street)
- taking a few puffs on a cigarette (each cigarette smoked reduces life expectancy by 10 minutes)
- an overweight person eating 20 extra calories (e.g., a quarter of a slice of bread and butter)
- driving an extra 5 miles in an automobile

These examples should put the risk of 1 mrem of radiation into proper perspective. Many more examples will be given in Chapter 8.

There has been intermittent publicity over the years about the fact that nuclear power plants, as a result of minor malfunctions or even in routine operation, occasionally release small amounts of radioactivity into the environment. As a result, people living very close to a plant receive about 1 mrem per year of extra radiation exposure. From the above example we see that, if moving away increases their commuting automobile travel by more than 5 miles per year (25 yards per day), or requires that they cross a street more than one extra time every 8 weeks, it is safer to live next to the nuclear plant, at least from the standpoint of routine radiation exposure.

SCIENTIFIC BASIS FOR RISK ESTIMATES

How do we arrive at the estimated cancer risk of low-level radiation, 1 chance in 4 million per millirem? We know a great deal about the cancer risk of high-level radiation, above 100,000 mrem, from various situations in which people were exposed to it and abnormally high cancer rates resulted.[11] The best-known example is the carefully followed group of 90,000 Japanese A-bomb survivors, among whom 8,500 people were exposed to doses in the range 100,000-600,000 mrem and suffered about 300 excess (i.e. more than would be normally expected) cancer deaths. In the period 1935-1954, it was fashionable in British medical circles to treat an arthritis of the spine called "ankylosing spondylitis" with heavy doses of X-rays, averaging about 300,000 mrem, which produced over 100 excess cancers among 14,000 patients so treated. In Germany, that disease and spinal tuberculosis were treated with radium injections which administered 900,000 mrem to the bone; in a study of 900 patients treated before 1952, there were 54 excess bone cancers.

The International Radiation Study of Cervical Cancer Patients has been following 182,000 women treated for cervical cancer with radiation in Canada, Denmark, Finland, Norway, Sweden, United States, United Kingdom, and Yugoslavia; exposures were typically 2-7 million mrem to the pelvis, 0.1-1 million mrem to the kidneys, stomach, pancreas, and liver; and lower doses to other body organs. There have been up to 250 excess cancers among them.

In Germany, Denmark, and Portugal, thorium (a naturally radioactive element) was injected into patients to aid in certain types of X-ray diagnosis between 1928 and 1955, giving several million millirem to the liver; among 3,000 patients there have been over 300 excess liver cancers. Between 1915 and 1935, numerals on luminous watch dials were hand painted in a New Jersey plant using a radium paint, and the tip of the brush was formed into a point with the tongue, thereby getting radium into the body; among 775 American women so employed, the average bone dose was 1.7 million mrem, and there were 48 excess bone cancers among them. (Over 3,000 other radium watch dial painters have been studied in less detail, with similar findings.)

There have been studies of about 10,000 radiologists from early in this century; there were numerous excess cancers from the very high doses they received, but unfortunately, it has been difficult to quantify those doses. There are data on about 10,000 patients treated with radiation for Hodgkins disease, among whom there was a large excess of leukemia and some excess of other types of cancer; however, chemotherapy, which also causes cancer, is an important complication here. There is information on patients treated with radiation for cancer of the ovaries, breast, and other organs who survived to develop other cancers from that radiation. There are data on patients given large doses of radiation for immune suppression in organ transplants (mainly for kidneys and bone marrow). Other studies include British women given X-ray treatments for a gynecologic malady, women in a Nova Scotia tuberculosis sanatorium who were repeatedly subjected to X-ray fluoroscopic examination, American women given localized X-ray treatment for inflammation of the breasts, American infants treated with X-rays for enlargement of the thymus gland and other problems, Israeli infants treated with X-rays for ringworm of the scalp, and natives of the Marshall Islands exposed to fallout from a nuclear bomb test.

Numerous studies have examined radon's effects on miners, since it tends to reach high concentrations in mines. Extensive studies have been carried out on uranium miners in the Colorado plateau, among whom there were 256 cases of lung cancer versus 59 expected cases; for uranium miners in Czechoslovakia, where there were 212 cases versus 40 expected; for iron miners in Sweden, who suffered 51 cases versus 15 expected, for uranium miners in Ontario, Canada, who experienced 87 cases versus 57 expected; and for several other smaller groups.

The most recent analyses of the data other than for radon are those reported in 1990 by the National Academy of Sciences Committee on Biological Effects of Ionizing Radiation (BEIR)[11] and in 1988 by the United Nations Scientific Committee on Effects of Atomic Radiation (UNSCEAR).[12] Both of them depended primarily on the Japanese A-bomb survivors and, to a lesser extent, the British ankylosing spondylitis patients. UNSCEAR also used the cervical cancer patients, and the BEIR analysis utilized data on the fluoroscopy patients and those treated with X-rays for inflammation of the breasts, thymus gland enlargement, and ringworm of the scalp. The analyses for effects of radon were carried out separately by another BEIR committee,[14] by the U.S. National Council on Radiation Protection and Measurements (NCRP),[15] and by the International Commission on Radiological Protection (ICRP).[16]

In the earlier BEIR analysis[17] of 1980, there was good general agreement among results obtained from the most important data sets. But during the 1980s, very extensive new evaluations were carried out on the radiation doses received by the Japanese A-bomb survivors, including new estimates of the radiation emitted from the bombs and more careful consideration of how each individual was shielded from this radiation by building materials and the outer parts of their bodies. As a result, the risk estimates from the A-bomb survivor data increased substantially and are now about 1.5 times and 3 times higher, respectively, than those from the ankylosing spondylitis and cervical cancer patients. They are also 5 times higher than the risks obtained from radon exposure to miners. Since both the BEIR and UN-SCEAR analyses give very heavy weight to the A-bomb survivor data, the risk estimates they obtain are substantially higher than those given previously.

Another problem in deriving estimates arises from the fact that the younger members of the exposed groups are nearly all still alive, and one must make estimates of how many of them will eventually die of cancer from the exposure. This is done by use of mathematical models, and the results obtained can vary considerably with the model chosen.[18] The UNSCEAR Report gives results for two different models, but the 1990 BEIR Report uses only the model that yields the higher risk. This risk comes out to be 0.78 chances in a million (1 chance in 1.3 million) of a fatal cancer per millirem of exposure to the whole body.

Note that this risk is based on the high doses, 100,000-600,000 mrem, received by the most exposed A-bomb survivors during a few seconds following the bomb explosions. There is a large body of evidence indicating that the risk per millirem is much less at low doses, especially if the dose is received over an extended time period; i.e., at low dose rates. In essentially all situations we will be discussing, dose and dose rates are in this low range, in which there are no direct experimental data for deriving risk estimates.

If we were to use the above stated BEIR risk estimate at low doses, we would be assuming that there is a straight line relationship, independent of dose rate, between cancer risk and radiation dose. That is, for example, given that there is a 0.78 (78%) risk of cancer from exposure to 1,000,000 mrem, we would be assuming that there is a risk of 0.78/1,000,000 from exposure to 1 mrem. Let us review the evidence demonstrating that the actual risk at low dose and low dose rate is much less than predicted by this straight line relationship.[12]

The theories of how radiation induces cancer predict this reduced risk effect, and experiments on both human and mouse cells exposed to radiation and grown in culture exhibit it. Experiments on laboratory animals injected with radioactive materials clearly show the reduced risk at low dose, as do experiments on animals exposed to external radiation sources. The data on radium dial painters indicate this behavior with high statistical significance, even though it involves alpha particles, a type of radiation where the low dose reduction is expected to be least important. One-third as many thyroid cancers as would be expected from the straight line theory were later found in 35,000 patients treated for hypothyroidism with radioactive iodine. In all of these cases, the straight line theory predicts far more cancers at low doses than are actually observed. There is a lot of evidence that cancers arising from lower radiation levels take longer to develop, which implies that cancers from low radiation doses would often be delayed until after death from other causes and therefore never materialize.

In view of all this evidence, both UNSCEAR[12] and NCRP[19] estimate that risks at low dose and low dose rate are lower than those obtained from the straight line relationship by a factor of 2 to 10. For example, if 1 million mrem gives a cancer risk of 0.78, the risk from 1 mrem is not 0.78 chances in a million as stated previously, but only ½ to ⅒ of that (0.39 to 0.078 chances in a million). The 1980 BEIR Committee accepted the concept of reduced risk at low dose and used it in its estimates. The 1990 BEIR Committee acknowledges the effect but states that there is not enough information available to quantify it and, therefore, presents results ignoring it but with a footnote stating that these results should be reduced. As an intermediate between the factor of 2 and 10, in this chapter I use the 1990 BEIR results reduced by a factor of 3 for situations where dose and dose rates are low. The risk is then 0.26 chances in a million (1 chance in 4 million) for a fatal cancer per millirem of exposure to the whole body.

In Chapter 6, and to a small extent in Chapters 11 and 12 of this book, however, I quote results from a variety of sources, and it would be a difficult task to go back through each of them and make corrections for the different risk factors they use. I have, therefore, not made such corrections. For many situations discussed there, cancer risks should be approximately doubled. In no case would this make a qualitative change in the conclusions.

THE MEDIA AND RADIATION

We now turn to the question of why the public became so irrationally fearful of radiation. Probably the most important reason is the gross over-coverage of radiation stories by television, magazines, and newspapers. Constantly hearing stories about radiation as a hazard gave people the sub-conscious impression that it was something to worry about. In attempting to document this overcoverage, I obtained the number of entries in the New York Times Information Bank on various types of accidents and compared them with the number of fatalities per year caused by these accidents in the United States. I did this for the years 1974-1978 so as not to include the Three Mile Island accident, which generated more stories than usual. On an average, there were 120 entries per year on motor vehicle accidents, which kill 50,000 Americans each year; 50 entries per year on industrial accidents, which kill 12,000; and 20 entries per year on asphyxiation accidents, which kill 4,500; note that for these the number of entries, which represents roughly the amount of newspaper coverage, is approximately proportional to the death toll they cause. But for accidents involving radiation, there were something like 200 entries per year, in spite of there not having been a single fatality from a radiation accident for over a decade.

From all of the hundred or so highly publicized incidents discussed earlier in this chapter (with the exception of the Three Mile Island accident), the total radiation received by all people involved was not more than 10,000 mrem.[20] Since we expect only one cancer death from every 4 million mrem, there is much less than a 1% chance that there will ever be even a single fatality from all of those incidents taken together. On an average, each of these highly publicized incidents involved less than 1 chance in 10,000 of a single fatality, but for some reason they got more attention than other accidents that were killing an average of 300 Americans every day and seriously injuring 10 times that number. Surely, then, the amount of coverage of radiation incidents was grossly out of proportion to the true hazard.

Another problem, especially in TV coverage, was use of inflammatory language. We often heard about "deadly radiation" or "lethal radioactivity," referring to a hazard that hadn't claimed a single victim for over a decade, and had caused less than five deaths in American history.[21] But we never heard about "lethal electricity," although 1,200 Americans were dying each year from electrocution; or about "lethal natural gas," which was killing 500 annually with asphyxiation accidents.

A more important problem with TV stories about radiation was that they never quantified the risk. I can understand their not giving doses in millirem—that may have been too technical for their audience—but they could have easily compared exposures with natural radiation or medical X-rays. In the 1982 accident at the Rochester power plant, which was the top story on the network evening news for two days, wouldn't it have been useful to tell the public that no one received as much exposure from that accident as he or she was receiving every day from natural sources? This is not a new suggestion; similar comparisons had consistently been made by scientists for 35 years in information booklets, magazine articles, and interviews, but the TV people never used them.

Another reason for public misunderstanding of radiation was that the television reports portrayed it as something very new and highly mysterious. There is, of course, nothing new about radiation because natural radioactivity has always been present on Earth, showering humans with hundreds of times more radiation than they can ever expect to get from the nuclear power industry. The "mystery" label was equally unwarranted. As mentioned earlier, radiation effects are much better understood by scientists than those of air pollution, food additives, chemical pollutants in water, or just about any other agent of environmental concern. There are several reasons for this. Radiation is basically a much simpler phenomenon, with simple and well-understood mechanisms for interacting with matter, whereas air pollution and the others may have dozens or even hundreds of important components interacting in complex and poorly understood ways. Radiation is easy to measure and quantify, with relatively cheap and reliable instruments providing highly sensitive and accurate data, whereas instruments for measuring other environmental agents are generally rather expensive, often erratic in behavior, and relatively insensitive. And finally our knowledge of radiation health effects benefits from a $2 billion research effort extending over 50 years. More important than the total amount of money is the fact that research funding for radiation health effects has been fairly stable, thereby attracting good scientists to the field, allowing several successive generations of graduate students to be trained and excellent laboratory facilities to be developed.

Television gave wide publicity to any scientist or any scientific work indicating that radiation might be more dangerous than the usual estimates. A study by Mancuso and collaborators[22] was the best-known example of such work. In spite of the fact that it was universally rejected by the scientific community,[23] and completely ignored by BEIR, UNSCEAR, ICRP, and other such groups, it was frequently referred to in the media. I remember the meeting of the Health Physics Society in Minneapolis in 1978, where Mancuso and some of his critics were scheduled to speak. The TV cameras were set up well ahead of time, and they operated continuously as Mancuso presented his arguments. But when he finished and his critics began to speak, the TV equipment was disassembled and carried away.

Innumerable stories of this sort could be recited. As a result, it came to pass that the public's estimate of the cancer risk of 1 mrem of radiation (a strictly scientific question if there ever was one), was not that of the National Academy of Sciences Committee, not that of the United Nations Scientific Committee, and not that of the International Commission on Radiological Protection (all three of which agreed), but rather that of TV producers with no scientific education or experience, and possibly influenced by political prejudices.

It was my impression that TV people considered the official committees of scientific experts to be tools of the nuclear industry rather than objective experts. The facts don't support that attitude. The National Academy of Sciences is a nonprofit organization chartered by the U.S. Congress in 1863 to further knowledge and advise the government. It is composed of about a thousand of our nation's most distinguished researchers from all branches of science. It appointed the BEIR Committee and reviewed its work. The BEIR Committee itself was composed of about 21 American scientists well recognized in the scientific community as experts in radiation biology; 13 of them were university professors, with lifelong job security guaranteed by academic tenure. The United Nations Scientific Committee on Effects of Atomic Radiation (UNSCEAR) was made up of scientists from 20 nations from both sides of the iron curtain and the Third World. The countries with representation on the International Commission on Radiological Protection (ICRP) were similarly distributed, and the chairman was from Sweden.

To believe that such highly reputable scientists conspired to practice deceit seems absurd, if for no other reason than that it would be easy to prove that they had done so and the consequences to their scientific careers would be devastating. All of them had such reputations that they could easily obtain a variety of excellent and well-paying academic positions independent of government or industry financing, so they were not vulnerable to economic pressures.

But above all, they are human beings who have chosen careers in a field dedicated to protection of the health of their fellow human beings; in fact, many of them are M.D.'s who have foregone financially lucrative careers in medical practice to become research scientists. To believe that nearly all of these scientists were somehow involved in a sinister plot to deceive the public indeed challenges the imagination.

The Media Institute, a Washington-based organization, published an extensive study[24] of TV network evening news coverage of nuclear power issues during the 1970s. By far the most often quoted source of information was the group opposed to nuclear power, Union of Concerned Scientists, whose membership includes only about 0.1 percent of all scientists. The most widely quoted "nuclear expert" was Ralph Nader. During the month following the Three Mile Island accident, the only scientist among the 10 most frequently quoted sources was Ernest Sternglass, almost universally regarded in the scientific community as one of the least reliable of all scientists (see next section). During that time there were no pronuclear "outside experts" among the top 10 quoted sources. The only pronuclear sources in the top 10 were from the nuclear industry, who clearly had low public credibility, especially during that period.

For those who can't understand why television excessively covered and distorted information about the hazards of radiation, I believe it was because their primary concern is entertainment rather than education. One point in the ratings for the network evening news is worth $11 million per year in advertising revenue. In that atmosphere, what would happen to a TV producer who decided to concentrate on properly educating the public rather than entertaining it? As an illustration of the low priority the networks place on their educational function, I doubt if there are more than one or two Ph.D. level scientists in the full-time employ of any television network, in spite of the fact that they are the primary source of science education for the public. Even a strictly liberal arts college with no interest in training scientists typically has one Ph.D.-level scientist for every 200 students, whereas the networks have practically none for their 200 million students.

If TV producers took their role of educating the public seriously, they would have considered it their function to transmit scientific information from the scientific community to the public. But this they didn't do. They wanted to decide what to transmit, which means that they made judgments on scientific issues. When I brought this to their attention, they always said that the scientific community was split on the issue of dangers from radiation. By "split" they seemed to mean that there was at least one scientist disagreeing with the others. They didn't seem to recognize that a unanimous conclusion of a National Academy of Sciences Committee should be given more weight than the opinion of one individual scientist who is far outside the mainstream. Their position was that, since the scientific community was split, they had no way to find out what the scientific consensus was. To this I always proposed a simple solution: pick a few major universities of their choice, call and ask the operator for the department chairman or a professor in the field, and ask the question; after five such calls the consensus would be clear on almost any question, usually 5 to 0. The TV people never were willing to do this. My strong impression was that they weren't really interested in what scientists had concluded. They were only after a story that would arouse viewer interest. Clearly, a scare story about the dangers of radiation serves this purpose best.

A POLL OF RADIATION HEALTH SCIENTISTS

Here and elsewhere in communicating with the public, I try to represent the position of the great majority of radiation health scientists. My understanding of this position is derived from innumerable conversations, remarks, and innuendos, but the number of people encompassed by these is rather limited. In 1982, I became concerned that I had no real proof that I was properly representing the scientific community. I, therefore, decided to conduct a poll by mail.

The selection of the sample to be polled was done by generally approved random sampling techniques using membership lists from Health Physics Society and Radiation Research Society, the principal professional societies for radiation health scientists. Selections were restricted to those employed by universities, since they would be less likely to be influenced by questions of employment security and more likely to be in contact with research. Procedures were such that anonymity was guaranteed.

Questionnaires were sent to 310 people, and 211 were returned, a reasonable response for a survey of this type. The questions and responses are given in Fig. 1.

In the questionnaire that was sent out, the spaces above the lines were blank, to be used by the respondent to check or insert a number. In questions 1, 2, and 3 of Fig. 1, the numbers on the lines are the numbers of respondents that checked that choice. In question 4, the entries on the lines start with an E or an A to designate whether the person or organization is generally considered to be part of "the Establishment," or mainstream (E), or "anti-Establishment," or out of the mainstream (A). Following the E or A is the average of all responses (which were individually numbers from 0 to 100), and in parentheses is the number of respondents who replied to that item rather than leaving it blank. In the questionnaire as sent out, the Establishment Scientists were named; there was no indication of whether the person or group was "Establishment" or "anti-Establishment."

The results listed in Fig. 1 strongly confirmed my feeling that the involved scientists considered the public's fear of radiation to be greatly exaggerated and that television coverage greatly exaggerated the dangers of radiation. Question 4 confirmed my strong impression that they supported the Establishment, whose conclusions I always use, and rejected the anti-Establishment scientists and groups. Note that this was a secret poll with guaranteed anonymity for respondents.

The responses to Question 3 were really shocking to me because respondents were voting against their own best interests. In all normal circumstances, scientists will claim that their field should get more money. But in this poll, under the protection of anonymity, they said that they were already getting too much, considering how trivial a danger they were protecting against. This represents a truly admirable degree of honesty.

GENETIC EFFECTS OF RADIATION[11,12,25]

Other than inducing cancer, the most significant health impact of low-level radiation is causing inherited disabilities in later generations. These

1. In comparing the general public's fear of radiation with actual dangers of radiation, I would say that the public's fear is (check one):

<u>2</u> grossly less than realistic (i.e., not enough fear).
<u>9</u> substantially less than realistic.
<u>8</u> approximately realistic.
<u>18</u> slightly greater than realistic.

<u>104</u> substantially greater than realistic.
<u>70</u> grossly greater than realistic (i.e., too much fear).

2. The impressions created by television coverage of the dangers of radiation (check one):

<u>59</u> grossly exaggerate the danger.
<u>110</u> substantially exaggerate the danger.
<u>26</u> slightly exaggerate the danger.
<u>5</u> are approximately correct.

<u>3</u> slightly underplay the danger.
<u>2</u> substantially underplay the danger.
<u>1</u> grossly underplay the danger.

3. From the standpoint of our national welfare and in comparison with other health threats from which the public needs protection, the amount of money now being spent on radiation protection in the United States is (check one):

<u>18</u> grossly excessive.
<u>35</u> substantially excessive.
<u>30</u> slightly excessive.
<u>62</u> about right.

<u>22</u> slightly insufficient.
<u>21</u> substantially insufficient.
<u>4</u> grossly insufficient.

4. How would you rate the scientific credibility in the field of radiation health of the following scientists or groups of scientists. Write a number between 0 and 100 indicating in which percentile of credibility of radiation health scientists their credibility falls; for example, "60" would mean that person's work is more credible than that of 60% of all scientists in the field, and less credible than that of 40%.

<u>E-82 (175)</u>	BEIR Committee (National Academy of Sciences)	<u>E-85 (157)</u>	National Council on Radiation Protection (NCRP)
<u>E-81 (116)</u>	UNSCEAR Committee (United Nations Scientific Commission)	<u>E-72 (60)</u>	Reactor Safety Study Panel on Health Effects
<u>A-25 (157)</u>	The Ralph Nader research organizations	<u>E-78 (102)</u>	Establishment Scientist A
<u>E-85 (157)</u>	International Commission on Radiological Protection (ICRP)	<u>A-14 (154)</u>	Ernest Sternglass
		<u>E-71 (113)</u>	Establishment Scientist B
		<u>A-30 (129)</u>	John Gofman
		<u>E-75 (52)</u>	Establishment Scientist C
<u>A-41 (122)</u>	Union of Concerned Scientists	<u>A-28 (96)</u>	Thomas Mancuso
		<u>E-80 (114)</u>	Establishment Scientist D
		<u>A-32 (60)</u>	Thomas Najarian
		<u>E-71 (80)</u>	Establishment Scientist E

Fig. 1. Questions and responses in poll of radiation health scientists (see text for explanation). Data from B.L. Cohen, "A Poll of Radiation Health Scientists," *Health Physics*, 50, 639 (1986).

disabilities, often called *genetic defects,* range from minor problems like color blindness to very serious maladies like mongolism. There is further discussion of their nature in the Chapter 5 Appendix.

Some people believe that radiation can produce two-headed children, or various types of subhuman or superhuman monsters. In order to understand why this cannot be so, one need only recognize that when a reproductive cell is struck by a particle of radiation, it has no way to "know" whether that radiation came from something produced in a nuclear reactor or from a natural source. The effects must, therefore, be the same. Since humans have always been exposed to natural radiation at intensities hundreds of times higher than what they will ever receive from the nuclear industry, no new types of genetic disease can be expected from the latter. In fact, if by some odd circumstance a new type of genetic disease were to develop, there would be no way to determine its cause, but it would be hundreds of times more likely to have been caused by natural radiation (or by medical X-rays) than by radiation from the nuclear industry. It would most likely be due to a spontaneous mutation.

Some people have the impression that radiation-induced genetic effects can destroy the human race, but that is also false. To understand this, consider the fact that new mutations are occurring all the time, in most cases spontaneously. How come the incidence of genetic disease is not continuously increasing? The reason lies in Darwin's famous law of natural selection: bad mutations reduce chances for success in reproduction and survival, and are therefore lost in the breeding process. As a very obvious example, children born with Down's syndrome almost never grow to adulthood and have offspring. Conversely, in the extremely rare situation where a mutation causes a favorable trait, it leads to increased success in reproduction and survival. Thus the law of natural selection causes good traits to be bred in and bad traits to be bred out of the human race, so any bad traits induced by radiation will eventually disappear. In the very long term, small additional radiation exposures can therefore only improve the human race. That is how evolution works. However, that effect is negligibly small. The important genetic effects are the short-term human misery created by genetic disease.

Since the radiation we can expect to receive from the nuclear industry is less than 1% of what we receive from natural sources, the genetic effects of the nuclear industry increase man's radiation exposure by less than 1% over that from natural radiation. As natural radiation is responsible for only about 3% of all normally encountered genetic defects,[11] the impact of a large-scale nuclear industry would increase the frequency of genetic defects by less than 1% of 3%, or less than 1 part in 3,000.

Another way to understand the genetic effects of nuclear power is to make a comparison with other human activities that induce genetic defects. One good example is older adults having children, which increases the likelihood of several types of genetic disease. Increased maternal age is known[26] to enhance the risk of Down's syndrome, Turner's syndrome, and several other chromosomal disorders, while increased paternal age rapidly raises the risk of achondroplasia (short-limbed dwarfism) and presumably a thousand other autosomal dominant diseases.[27] The genetic effects of a large nuclear industry would be equal to those of delaying the conception of children by an average of 2.6 days.[25] Between 1960 and 1973, the average age of parenthood increased by about 50 days, causing 20 times as much genetic disease as would be induced by a large nuclear industry.

To be quantitative about genetic effects of radiation, we can expect one genetic defect in all future generations combined for every 11 million mrem of individual radiation exposure to the general population. For example, natural radiation exposes the gonads to 85 mrem per year, or a total of $(85 \times 240$ million $=)$ 21 billion mrem per year to the whole U.S. population. It can therefore be expected to cause (21 billion/11 million $=)$ 2,000 cases per year of genetic disease, about 2% of all the cases normally occurring. (Since our present population includes an abnormally large number of younger people, this result is increased to 3%.) This is the estimate of the National Academy of Sciences BEIR Committee,[11] and those of UN-SCEAR[12] and ICRP[28] are similar. It is interesting to point out that these estimates are derived from studies on mice, because there is no actual evidence for radiation causing genetic disease in humans. The best possibility for finding such evidence is among the survivors of the A-bomb attacks on Japan, but several careful studies have found no evidence for an excess of genetic defects among the first generation of children born to them. If humans were appreciably more susceptible to genetic disease than mice, a clear excess would have been found. We can therefore be confident that in utilizing data on mice to estimate effects on humans, we are not understating the risk.

Often an individual worries about his or her own personal risk of having a genetically defective child; it is about 1 chance in 40 million for each millirem of exposure received prior to conception — somewhat more for men and somewhat less for women. This is equal to the risk of delaying conception by 1.2 hours.[25]

It may be relevant here to mention that air pollution and a number of chemicals can also cause genetic defects. There is at least some mutagenic information[29] on over 3,500 chemicals including bisulfites, which are formed when sulfur dioxide is dissolved in water (these have caused genetic changes in viruses, bacteria, and plants), and nitrosamines and nitrous acid, which can be formed from nitrogen oxides. Sulfur dioxide and nitrogen oxides are the two most important components of air pollution from coal burning. Other residues of coal burning known to cause genetic transformations are benzo-a-pyrene (evidence in viruses, fruit flies, and mice), ozone, and large families of compounds similar to these two. The genetic effects of chemicals on humans are not well understood, and there is practically no quantitative information on them, but there is no reason to believe that the genetic impacts of air pollution from coal burning are less harmful than those of nuclear power.

Caffeine and alcohol are known to cause genetic defects. One study[30] concludes that drinking one ounce of alcohol is genetically equivalent to 140 mrem of radiation exposure, and a cup of coffee is equivalent to 2.4 mrem.

Perhaps the most important human activity that causes genetic defects is the custom of men wearing pants.[31] This warms the sex cells and thereby increases the probability for spontaneous mutations, the principal source of genetic disease. Present very crude estimates are that the genetic effects of 1 mrem of radiation are equivalent to those of 5 hours of wearing pants.

TV coverage of genetic effects of radiation has been sparse, but there was one TV special which featured two beautiful twin babies (dressed in very cute dresses) afflicted with Hurler's syndrome, a devastating genetic disease. All sorts of details were offered on its horrors—they will go blind and deaf by the time they are 5 years old, and then suffer from problems with their hearts, lungs, livers, and kidneys, before they die at about the age of 10. Their father, who had worked with radiation for a short time, told the audience that he was sure that his radiation exposure had caused the genetic disease of his children. There was no mention of the fact that the father's total occupational radiation exposure was only 1,300 mrem, less than half of his exposure to natural radiation up to the time his children were conceived. With that much exposure, the risk of a child deriving a genetic defect is one chance in 25,000; their normal risk is 3%, due to spontaneous mutations, so there is only one chance in a thousand that their genetic problems were due to their father's job-related radiation exposure.

If one is especially concerned with genetic effects, there is much that can be done to reduce them. By using technology now available, like amniocentesis, sonography, and alpha fetoprotein quantification, we could avert 6,000 cases per year of genetic defects in the United States, at a total cost of

$160 million in medical services.[32] By comparison, a large nuclear power industry in the United States would eventually cause 37 cases per year, while producing a product worth $50 billion and paying about $8 billion in taxes. Thus, if 2% of this tax revenue—$160 million—were used to avert genetic disease, it could be said that the nuclear industry is averting $(6,000 \div 37 =)$ 160 cases for every case it causes. If the money were spent on genetic research, it would be even more effective.

One bothersome aspect of the genetic impacts of nuclear power is the conscience-burdening idea that we will be enjoying the benefits of the energy produced while future generations will be bearing the costs. However, we must recognize that there are many other, and much more important, situations in which our generation and its technology are adversely affecting the future. Perhaps the most important is our consumption of oil, gas, coal, and other mineral resources mentioned in Chapter 3; we are also overpopulating the world, exhausting agricultural land, developing destructive weapons systems, and cutting down forests. By comparison with any one of these activities, the genetic effects of radiation from the nuclear industry are exceedingly trivial. Moreover, this is not a new situation: forests in many parts of Europe were cut down at a time when wood seemed to be the most important resource for both energy and structural materials, and local exhaustion of agricultural land, game, and fish stocks has been going on for millennia.

Nevertheless, at least in recent times, each succeeding generation has lived longer, healthier, and more rewarding lives. The reason, of course, is that each new generation receives from its predecessors not only a legacy of detriments, but also a legacy of benefits. We leave to our progeny a tremendous fund of knowledge and understanding, material assets including roads, bridges, buildings, transport systems, and industrial facilities, well-organized and generally well-functioning political, economic, social, and educational institutions, and so on, all far surpassing what we received from our forbears. The important thing from an ethical standpoint is not that we leave our progeny no detrimental legacies—that would be completely unrealistic and counterproductive to all concerned—but rather that we leave them more beneficial than detrimental legacies.

In the case of the nuclear industry and the genetic effects it imposes on later generations, any meaningful evaluation must balance the value to future generations of an everlasting source of cheap and abundant energy developed at a cost of tens of billions of dollars and tens of thousands of man-years of effort, against a few cases of genetic disease which we also leave them the tools to combat cheaply and efficiently.

OTHER HEALTH IMPACTS OF RADIATION

When a particle of radiation penetrates a cell, the damage it does may cause the cell to die. If enough cells in a body organ die, the organ may cease to function, and this can lead to a person's death by what is termed radiation sickness. A dose of 500,000 mrem received over a short time period gives about a 50% risk of death, and with 1,000,000 mrem this risk is 100% unless there is heroic medical intervention, as by bone marrow transplants. After such an intense exposure, loss of hair, swelling, and vomiting are typical symptoms. If death does not occur within 30 days, the victim normally recovers fully.

There were a few deaths from radiation sickness in government operations, with the last one occurring in 1963, and there was at least one death from an error in a hospital. But this disease is relevant to nuclear power only in the most disastrous type of reactor accident. It will be discussed in the next chapter.

We might wonder whether diseases other than radiation sickness, cancer, and genetic defects can result from radiation. Careful studies[33] among the survivors of the atomic bomb attacks on Japan revealed no such evidence. Moreover, our understanding of how various diseases develop leads us not to expect other diseases from radiation.

One other effect of radiation is developmental abnormalities among children exposed to radiation *in utero*.[11,34] This is well known from animal studies, and there is also extensive human evidence from medical exposures and from studies of the Japanese A-bomb survivors. Among the latter, children exposed prior to birth to more than 25,000 mrem were, at age 17, an average of 0.9 inches shorter, 7 pounds lighter, and nearly a half inch smaller in head diameter than average. Among the 22 children who received more than 150,000 mrem from the bomb before the 18th week of gestation, 13 had small head size and 8 suffered from mental retardation. There were only two cases of mental retardation among those exposed after the 18th week. There was no evidence for mental retardation for exposures less than 50,000 mrem. From animal studies, it is expected that there may be slight developmental abnormalities for doses as low as a few thousand millirem at critical times during fetal development. (A woman living very close to a nuclear power plant would typically receive less than 1 mrem during pregnancy.)

Researchers have also investigated the possibility that *in utero* exposure may give a large risk of childhood cancer.[34] It was initially reported that children whose mothers received pelvic X-rays during pregnancy had ten-fold elevated risks for this disease. However, many factors other than X-rays have been found to be similarly correlated with childhood cancer, including use by the mother of aspirin and of cold tablets, and the child's blood type, viral infections, allergies, and appetite for fish and chips (this was a British study). It was shown that the original correlations would have predicted 18 excess childhood cancers among those exposed *in utero* to the A-bomb attacks on Japan, whereas none occurred. The initial observations could be explained as effects of the medical problems that required the mothers' X-rays. A study of children whose mothers were given X-rays for nonmedical reasons showed no evidence for increased cancer. Nevertheless, the official committees still consider this excess cancer risk to be a possibility; it is therefore taken into account in setting radiation protection regulations for occupational exposure of pregnant (or potentially pregnant) women. From the public health viewpoint, *in utero* exposure is not of great importance since only 1% of the population is *in utero* at any given time.

These *in utero* effects are often exaggerated in the public mind. There were widely circulated pictures of grossly deformed farm animals with claims that they were caused by *in utero* exposure to radiation from the Three Mile Island accident. There were no such claims to be found in the scientific literature. Recall that average exposures from that accident were 1.2 mrem, which is equal to the radiation received from natural sources every 5 days, so even if the deformities were due to radiation, they would much more likely be due to natural radiation. There have been large numbers of careful experiments on the effects of radiation to various animals—I have visited farms operated for that purpose in Tennessee and in Idaho—and no such effects were observed in those experiments from exposures less than several thousand millirem. It is therefore extremely hard to believe that such effects can occur at 1.2 mrem. Naturally, a small fraction of all animals (or humans) born anywhere and under any circumstances are deformed, so there would be no great difficulty in collecting pictures of deformed animals. Before one can claim evidence that these effects are due to radiation from the Three Mile Island accident, however, it would have to be shown that the number of deformed animals in that area at that time was larger than the number in that area at previous times, or in other areas at that time, by a statistically significant amount. That has certainly never been shown for any health effects among humans or animals.

IRRATIONAL FEAR OF RADIATION

Because of the factors we have been discussing and perhaps some others, the public has become irrational over fear of radiation. Its understanding of radiation dangers has virtually lost all contact with the actual dangers as understood by scientists. Perhaps the best example of this was the howl of public protest when plans were announced more than a year after the accident at Three Mile Island to release the radioactive gas that had been sealed inside the containment structure of the damaged reactor. This was important so that some of the safety systems could be serviced, and it was obviously necessary before recovery work could begin. Releasing this gas would expose no one to as much as 1 mrem, and the exposure to most of the protesters would be a hundred times less. Simply traveling to a protest meeting exposed the attenders to far more danger than release of the gas; moreover, an appreciable number fled the area, traveling a hundred miles or more, at the time of the release. Recall that 1 mrem of radiation has the same risk as driving 5 miles or crossing a street five times on foot. Needless to say, the statements of fear by the protesters were transmitted to the national TV audience with no accompanying evidence that their fears were irrational.

One disheartening aspect of that episode was the effort by the Nuclear Regulatory Commission (NRC) to handle it. An early survey of the local citizenry revealed that there was substantial fear of the release of the gas. The NRC therefore undertook a large program of public education, explaining how trivial the health risks were. When this public education campaign was completed, another poll of the local citizenry was taken. It showed that the public's fear was greater than it was before the campaign. The public's reaction on matters of radiation defied all rational explanation.

One tragic consequence of this public irrationality was the impact it had on medical uses of radiation. Radioactive materials and accelerator radiation sources are widely used for medical diagnosis and therapy, saving tens of thousands of lives each year. Even those opposed to nuclear energy claim that there may be an average of only a few hundred lives lost per year from all aspects of nuclear power (see Chapter 6), whereas most estimates are more than 10 times lower. Thus manmade radiation is saving hundreds or thousands of times as many lives as it is destroying. But as a result of the public's irrationality, patients were refusing radiation procedures with growing frequency, and physicians were becoming more hesitant to use them. In less than an hour of discussion during intermissions at a local meeting on nuclear medicine in New York City, I learned of the following situations:

- In one big-city hospital, about 20% of the patients refuse the recommended treatment for hyperthyroidism, which involves the use of radioactive iodine, opting instead for a less cost-effective drug treatment that frequently leads to relapse.
- In another hospital nursery intensive care unit, when a portable X-ray

machine is brought into the unit, the nurses leave the area, abandoning the infants under their care; it is estimated that remaining at their posts would expose them to less than 1 mrem per year of additional radiation.

- A large medical center planned a project involving radiation for 100 patients per year in its Intensive Care Unit. It was estimated that the 10 nurses working there would each be exposed to about 100 mrem per year of additional radiation, approximately the extra amount received by living in Colorado. The nurses threatened to strike, forcing abandonment of the project.

I don't know of any elaborate quantitative estimates of the percentage by which fear is reducing medical use of radiation, but even if it is only 10%, the public's irrational fear of radiation is killing at least many hundred, and probably several thousand, Americans each year.

SUMMARY

We have shown that any discussion of radiation hazards must be quantitative, giving doses in millirem. If we know the radiation dose in millirem, we can estimate the risk. We can then understand this risk by comparing it with other risks with which we are familiar. Unfortunately, the public does not go through this rational process. Largely because of poor television coverage, the public has reacted irrationally and emotionally to even the most trivial radiation hazards. A lot of education is needed here, and hopefully this chapter will contribute to that process. But in any case, the risk estimates described here will be used extensively in the rest of this book.

Chapter 6 / THE FEARSOME REACTOR MELTDOWN ACCIDENT

Technologies are normally developed by entrepreneurs whose primary goal is making money. If the technology is successful, the entrepreneurs prosper as a new industry develops and thrives. In the process, the environmental impacts of this new technology are the least of their concerns. Only after the public revolts against the pollution inflicted upon it does the issue of the environment come into the picture. At that point an adversarial relationship may develop, with the government serving to protect the public at the expense of the industry. Coal-burning technologies have been an excellent example of this development process.

With nuclear energy, everything was to be entirely different. It was conceived and brought into being by the world's greatest scientists. They banded together to obtain government support; the highly publicized letter from Albert Einstein to President Roosevelt in 1941 was a key element in that process. Their motivation was entirely idealistic. None of them thought about making money, and there was no mechanism for them to do so. Their first objective was to save the world from the hideous Hitler, and after World War II it was to protect freedom and democracy through military strength. But from the beginning of the project in the early 1940s, the scientists always felt strongly that this new technology, developed at government expense, would provide great benefits to mankind.

Distinguished scientists like Henry Smythe and Glenn Seaborg held high positions of power all the way up until the early 1970s, and through them many of the greatest and most idealistic scientists, like Enrico Fermi, Eugene Wigner, and Hans Bethe, exerted great influence on the course of events. Directly or indirectly, hundreds of scientists were involved in guiding our national nuclear energy program. They set up national laboratories of unprecedented size in the New York (Brookhaven), Chicago (Argonne), and San Francisco (Berkeley, Livermore) areas and at the wartime development sites in Oak Ridge, Tennessee and Los Alamos, New Mexico. They arranged for an unprecedented level of financial support for research in universities where most of the scientists were based. Their objectives went far beyond development of nuclear technology, and included seeking a thorough understanding of the environmental effects. Their approach ran the gamut from the most basic research to the most practical applications.

The government's side of this enterprise was run by the Atomic Energy Commission (AEC). The AEC was set up at the behest of scientists to remove nuclear energy development from military control. Prominent scientists served as commissioners, often as chairmen. It's General Advisory Committee, made up of some of the nation's most distinguished scientists, exerted very strong influence. The AEC was monitored by the Congressional Joint Committee on Atomic Energy, which included some of the most powerful senators and representatives. A spirit of close cooperation reigned throughout. The goal of all was to provide humankind with the blessings of nuclear energy as expeditiously as possible.

There was a general understanding among all concerned that the scientists had paid their dues—they had given the government's military nuclear weapons, nuclear submarines, and a host of other goodies—and that their new technology was to serve humanity under the guidance of this research enterprise. Government also recognized that this enterprise, in the long run, would serve the public interest, and continues to support it to this day. A recent well-publicized element of that program is the multibillion dollar superconducting supercollider accelerator to be constructed in Texas to study the fundamental nature of matter, with no practical applications in sight.

Use of nuclear energy to generate electricity was a very important part of this research and development program. In order to promote it, the AEC brought in commercial interests beginning in the mid 1950s, but it kept the national laboratories deeply involved. One of the highest priority activities it assigned to them was investigation of the environmental impacts of nuclear power. For the first time in history, environmental impacts were thoroughly investigated *before* an industry started.

An important part of this effort was to try to "dream up" anything and everything that can possibly go wrong in a nuclear power plant, and investigate the consequences. This was a useful process in deciding on what safety systems to include. But if enough thought and "dreaming up" is devoted to any system, one can always devise a chain of events that can defeat all safety systems and do harm to the workers or the public. Though this is true for every technology, no other technology has ever been subjected to this degree of scrutiny.

These efforts to evaluate the risks of nuclear power plant accidents have been a valuable and successful scientific and engineering undertaking. Researchers broke new ground in the science of risk analysis (their developments are now being applied in other technologies). Results of this research have suggested new areas for investigation of nuclear reactor safety questions, and looking into these areas has been productive. This wide-ranging research program developed a variety of accident scenarios, calculated their consequences, and estimated their probability for occurring.

However, while these efforts were highly laudable, their effects proved to be disastrous. The public did not understand these risk analyses. Its attention became entirely focussed on excerpts stating that nuclear accidents can kill tens of thousands of people. They never seemed to notice that these reports estimated that such an accident can be expected only once in 10 million years. The public doesn't understand probabilities anyhow. Most people recognize little difference between a risk with a probability of once in 10 thousand years and once in 10 million years. The common impression was that a reactor meltdown accident killing tens of thousands of people would occur every few years.

In 1978, a movie called "The China Syndrome," based on this sort of thinking and starring some of Hollywood's top performers, gained widespread popularity. When the Three Mile Island accident followed in 1979, it became the news media story of the decade, complete with days of suspense during which the public was led to believe that a horrible disaster could occur at any moment. This combination of events led to very serious problems for the nuclear power industry.

As a result of these developments, the word *meltdown* has become a household word. We will use it here, although it is no longer used by risk analysis scientists. In the mind of the public, it refers to an accident in which all of the fuel becomes so hot that it forms a molten mass which melts its way through the reactor vessel. Let's use the word in that sense. The media frequently referred to it as "the ultimate disaster," evoking images of stacks of dead bodies amid a devastated landscape, much like the aftermath of a nuclear bomb attack.

On the other hand, the authors of the two principal reports on the Three Mile Island accident[1,2] agree that even if there had been a complete meltdown in that reactor, there very probably would have been essentially no harm to human health and no environmental damage. I know of no technical reports that have claimed otherwise. Moreover, all scientific studies agree that in the great majority of meltdown accidents there would be no detectable effects on human health, immediately or in later years. According to the government estimate, a meltdown would have to occur every week or so somewhere in the United States before nuclear power would be as dangerous as coal burning.

Even the Chernobyl accident, which was worse in many ways than any meltdown that can be envisioned for an American reactor, caused no injuries outside the plant. That is not to say that it is impossible to have fatalities caused by a meltdown, but it is estimated that in no more than 1 in 100 meltdowns could any be obviously related to the accident.

WAS THE THREE MILE ISLAND ACCIDENT A NEAR MISS TO DISASTER?

One of the principal reasons for the discrepancy between the public's impressions and the technical analyses is that nuclear reactors are sealed inside a very powerfully built structure called the "containment." Under ordinary circumstances the containment would prevent the escape of radioactivity even if the reactor fuel were to melt completely and escape from the reactor vessel. A typical containment[3,4,5] is constructed of 3-foot-thick concrete walls heavily reinforced by thick steel rods (see Fig. 1) welded into a tight net around which the concrete is poured. In fact, there is so much steel reinforcing that special techniques had to be developed to get the concrete to become distributed around it as it is poured. In addition, the inside of the containment is lined with thick steel plate welded to form a tight chamber which can withstand very high internal pressure, as high as 10 times normal atmospheric pressure.

The containment provides a broad range of protection for the reactor

Fig. 1. Construction workers on the steel-rod-reinforced containment structure of a Westinghouse reactor. Note the thickness and density of the reinforcing rods.

against external forces, such as a tornado hurling an automobile, a tree, or a house against it, an airplane flying into it, or a large charge of chemical explosive detonated against it. In a meltdown accident, however, the function of the containment is to hold the radioactive material inside. Actually, it need only do this for several hours, because there are systems inside the containment for removing the radioactivity from the atmosphere. One type blows the air through filters in an operation similar in principle to that of household vacuum cleaners. In another, water sprinklers remove the dust from the air. There are charcoal filter beds or chemical sprays for removing certain types of airborne radioactivity. Most radioactive materials, however, would simply get stuck to the walls of the building and the equipment inside, and thereby be removed from the air. Thus, if the containment holds even for several hours, the health consequences of a meltdown would be greatly mitigated. In the Three Mile Island accident, there was no threat to the containment. The investigations have therefore concluded that even if there had been a complete meltdown and the molten fuel had escaped from the reactor, the containment would very probably have prevented the escape of any large amount of radioactivity.[1,2] In other words, even if the Three Mile Island accident was a "near miss" to a complete meltdown (a highly debatable point), it was definitely *not* a near miss to a health disaster.

The Chernobyl reactor did not have a containment anything like those used in U.S. reactors. Analyses have shown, that if it had used one, virtually no radioactivity would have escaped, there would have been no threat to human health, and the world would probably have never heard about it.

ROADS TO MELTDOWN

In order to understand the meltdown accident, we must go back to its origins. A nuclear power reactor is basically just a water heater, evolving heat from fission processes in the fuel. This heats the water surrounding the fuel (see Fig. 2), and the hot water is used to produce steam. The steam is then employed as in coal- or oil-fired power plants to drive a turbine which turns a generator (sometimes called a "dynamo") which produces electric power (see Fig. 3). There are two different types of reactors in widespread use in the United States, pressurized water reactors (PWRs)[4] and boiling water reactors (BWRs)[5]. In the PWR, the heated water is pumped out of the reactor to separate units called "steam generators," where the heat in this water is used to produce steam. In the BWR, the steam is produced directly in the reactor so there is no need for a steam generator.

There are features of the nuclear water heater that differentiate it from water heaters in our basements or the coal- or oil-fired boilers that produce steam for various purposes in industrial plants. First, the waste products from the burning do not go up a chimney or settle to the bottom as an ash, but rather are retained inside the fuel. Nuclear fuel does not crumble into ashes

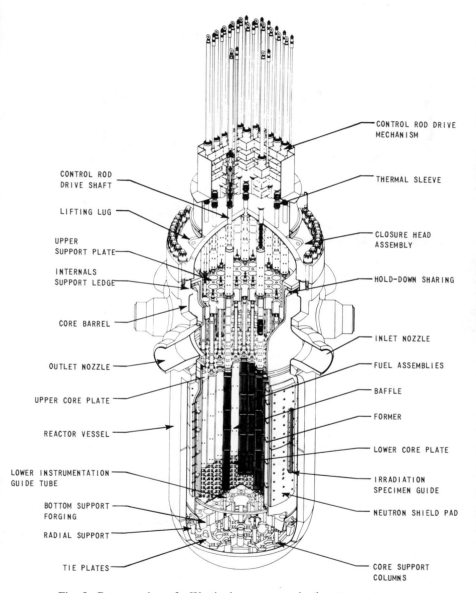

Fig. 2. Cutaway view of a Westinghouse pressurized water reactor.

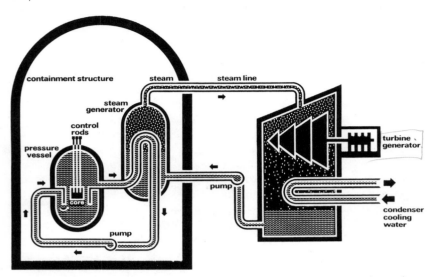

Fig. 3. Diagram (highly simplified) of a pressurized water reactor power plant. Water is heated to 6000°F by energy released in fission reactions in the reactor (it is prevented from boiling by maintaining high pressure), and pumped into the steam generator, where its heat is transferred to a secondary water system. The water in the latter is thereby boiled to make steam, which drives the turbines. The turbine drives the generator which produces electricity. It is necessary to condense the steam into water, greatly reducing its pressure, after it exits the turbine—otherwise there would be no tendency for the steam to rush through the turbine and thereby cause it to turn. The steam is condensed in the condenser by cooling it with water brought in from some outside source. The water formed by condensation is pumped back into the steam generator to be reused.

or get converted into a gas when burned, as do coal and oil fuels. Second, these waste products are radioactive, which means that they emit radiation. Third, because of their radioactivity, these wastes continue to heat the fuel even after the reactor is shut down[6]; it is therefore necessary to continue to provide some water to carry this heat away.

If, for some reason, no water is available to remove this heat (called a loss-of-coolant accident, LOCA), the fuel will heat up and eventually melt.[7] Fuel melting releases the radioactivity sealed inside. Some of this radioactivity would come off as airborne dust that has a potential for damaging public health if it is released into the environment. If there is some water in the reactor but not enough, the situation may be even worse, because steam reacts chemically with the fuel-casing material (an alloy of zirconium) at high temperature (2,700°F), releasing hydrogen, an inflammable and potentially explosive gas, and providing additional heat, thereby accelerating the fuel-melting process.

In the Three Mile Island accident,[8] the LOCA occurred as a result of a valve failing to close, while the operators were led to believe that it was closed; they had misinterpreted the information available to them from instrument readings. According to one estimate,[2] a complete fuel meltdown might have occurred if the water had continued to escape through the open valve for another 30 to 60 minutes.

How close was Three Mile Island to such a complete meltdown? There were many unusual aspects to the instrument readings at the time. Clearly, something very strange was going on. A number of knowledgeable people were trying to figure out what to do. One rightfully suggested closing an auxiliary valve in the pipe through which water was escaping. Within less than a minute after it was closed, a telephone call came in from another expert working at home suggesting that this auxiliary valve be closed,[2] so it cannot be claimed that a meltdown was prevented by the luck of one man's recognizing the right thing to do. It is difficult to prove that if neither of the two had thought of closing the valve someone else would have, but there were a lot of people involved in analyzing the information, and there would have been further clues developing before a meltdown would have occurred. Some analyses indicate that there would not have been a complete meltdown even if the valve had not been closed, as there was a small amount of water still being pumped in.

In any case, the widely publicized statement that the Three Mile Island accident came within 30 to 60 minutes of a meltdown seemed to be sufficient to scare the public. I often wonder why this is so—when we drive on a high-speed highway, on every curve we are within a few seconds of being killed if nothing is done—that is, if the steering wheel is not turned at the proper time. And don't forget that even if a meltdown had occurred, there very probably would have been no health consequences, since the radioactivity would have been contained.

As a result of the Three Mile Island accident[9] great improvements have been made in instrumentation, information availability to the operators, and operator training. There is now a requirement that a graduate engineer be on hand at all times. There will probably never again be a LOCA arising from faulty interpretation of instrument readings.

With that road to a meltdown now essentially blocked, let us consider what are believed to be the most probable roads now open.

1. *LOCA arising from a break in the reactor coolant system*. The cool-

ing water system for transferring heat out of the reactor operates at very high temperature and pressure (600°F, and 2,200 pounds per square inch [psi] in a PWR, or 1,200 psi in a BWR). Therefore, if the system should break open, the water would come shooting out as steam in a process picturesquely called *blowdown*. Such a break could arise from a failure in the seal of the huge pump that brings the water into the reactor, or from a pressure relief valve opened by a brief pressure surge failing to close, but the most likely cause would be a pipe breaking off, especially at a welded joint.

A series of safety measures is designed to protect the system from breaking open.[10] The first of these is very elaborate quality control on materials and workmanship, far superior to that in any other industry. No effort or expense is spared in choosing the highest quality materials and equipment, nor in requiring the most demanding specifications for safety-related parts of the system. The second measure is a highly elaborate inspection program, including X-ray inspection of every weld, and other inspections with magnetic particle and ultrasonic techniques during construction, followed by periodic ultrasonic and visual inspections after the reactor has gone into operation. The visual inspection program, for example, includes removal of insulation from pipes to search for imperfections or signs of cracking. One problem originally discovered by these inspections, "corrosion cracking," is discussed in the last section of this chapter. A third measure is a variety of leak detection systems: ordinarily a large break starts out as a small crack which allows some of the water and the radioactivity it contains to leak out. Leaking water becomes steam as it emerges (its temperature is close to 600°F), increasing the humidity; there are instruments installed to detect this increased humidity. Much of the radioactive material emerging with the leaking water attaches to airborne dust, and there are instruments in place for detecting increased radioactivity in this dust. These systems for detecting increased humidity and increased radioactivity in dust act as sensitive indicators for leaks, therefore serving as early warnings of possible cracks in the system.

If all of these measures should fail, a LOCA would occur. The remaining protection against a meltdown would then be the emergency core cooling system to be discussed below.

2. *Loss of electric power (station blackout).* If there should be no electric power to operate the pumps, the water in the reactor would stay there and get hotter and hotter, building up the pressure until relief valves open allowing the water to escape. In addition, the pump seals require cooling water and would fail if it were not supplied due to station blackout, leading to a LOCA as described in (1).

To protect against station blackout, off-site power is normally brought into the plant from two different directions, and several diesel-driven generators are available, any one of which could provide the needed power. These engines are started at frequent intervals to assure their availability when needed, and statistics are kept on their failures to start. In addition, some safety system pumps are driven by steam from the reactor rather than by electric power; normally there would be plenty of this steam available. Control systems for operating pumps and valves are electrically operated, but batteries are available for this purpose.

In some plants, at least, being without electric power for more than about 20 minutes would usually lead to a meltdown.

3. *Transients with failure of reactor protection systems.* While a reactor is operating, changes can and do occur which tend to increase the power level of a reactor. For example, the temperature or pressure of the water or its chemical content may change, causing it to absorb fewer neutrons, leaving more neutrons to strike uranium atoms and thus produce more energy. Reactors have "control rods," simple rods made of a material that strongly absorbs neutrons, which are moved in or out of the reactor core to control the power level. In the above example they would be moved in a short distance to absorb more neutrons and thus compensate for the fact that the water was absorbing fewer neutrons, restoring the power to its original level. An incident of this type is called a "transient." Small transients occur frequently in normal reactor operation, and control rods are frequently moved to adjust the power of the reactor.

Occasionally, perhaps once or twice a year, an abnormally large transient occurs which cannot be accommodated by the normal control rods. For example, if the electric power demand should suddenly drop drastically as in the case of a transformer or transmission line failure, the reactor is suddenly in a condition where it is producing far too much heat. For such transients, anticipated to occur many times in a reactor's "lifetime," safety systems automatically insert emergency control rods all the way into the reactor at high speed, absorbing so many neutrons that the chain reaction is completely stopped. This process is called *scram.*

It is possible that the scram system might fail when one of these anticipated large transients occur. This is called "anticipated transients without scram," or ATWS. An ATWS event would lead to rapid, intense overheating and loss of water—blown out through pressure relief valves. This loss of water would constitute a LOCA. It would also stop the chain reaction.

The protection against occurrence of an ATWS accident is in the use of high-quality materials and components, and in a good program of inspections and tests. If an ATWS does occur, the emergency core cooling system, to be discussed below, would normally prevent a meltdown.

4. *Earthquakes and fires.* Earthquakes can cause any of the above failures and can cause failures in safety systems which would ordinarily mitigate the effects of these failures. Fires, especially in the switch gear, in the control room, or in cables, can lead to failure of various operating or protection systems. For these reasons, nuclear plants are constructed with several features, like bracing and special pipe supports, to minimize effects of earthquakes. In addition, great care is taken in siting plants to avoid proximity to potentially active geological faults. (Widely circulated stories about plants being built on faults are not true.) Some of the best earthquake scientists in the nation are involved in this activity, and regulations and procedures are very elaborate.

Any system can be destroyed by a sufficiently powerful earthquake, but in an earthquake strong enough to cause a nuclear reactor meltdown, effects of the meltdown would be a relatively minor addition to the consequences of that earthquake.

All of these accident scenarios lead to loss of water. The chain reaction cannot go on without water, so it is shut down, but one must still worry about heat from radioactivity causing the fuel to melt. This can only be prevented if water cooling is very rapidly restored to the reactor core (where the fuel is located). Reactor designs provide this function through the "emergency core cooling system," ECCS. An ECCS consists of several independent systems for pumping water into the reactor, any one of which would provide sufficient water to save the reactor in most cases—in all cases two would do the job.[11] More details are given on the ECCS in the Chapter 6 Appendix.

Without water cooling, reactor fuel heats up very rapidly, and it would require perhaps 30 seconds before water from the ECCS would flood the reactor vessel to a level at which the core is covered. During this time, the fuel would reach temperatures in the range 1,000-2,500°F.

When the water from the ECCS first reaches the hot fuel, it would flash into steam, and at one time there was some concern as to whether this might prevent further water from reaching and cooling the fuel. Some of the first tests of small mock-ups, performed in 1970-1971, indicated that this might be the case. The problem thus received very wide publicity.[12] This was the situation that brought the opposition group, Union of Concerned Scientists (UCS), into prominence, as they asked for a halt to reactor licensing until the problem was resolved.[13] The culmination was a series of hearings held in Washington extending over a year in 1972-1973.

As a result of the hearings, changes were introduced in reactor operation as a temporary measure to reduce the performance required of the ECCS if a LOCA should occur, and a crash research program costing hundreds of millions of dollars was instigated to settle the unresolved questions. As more sophisticated experimental tests and computer analyses were developed, it became increasingly clear in the 1975-1978 time period that the ECCS would work. There were over 50 tests, far more realistic and sophisticated than the 1971 tests, and all came out favorably. The question was finally resolved in 1978 when a test reactor specifically designed to test the ECCS (called LOFT, for loss of fluid test) came into operation at the Idaho Nuclear Engineering Laboratory and was put through various types of LOCAs. In all cases, the ECCS performed better than had been estimated.[14] For example, in the first LOFT test, the best estimate from the computer analysis was a maximum temperature of 1,376°F, the conservative calculation used for the safety analysis gave 2,018°F, but the highest measured temperature was only 960°F. In the second LOFT test, carried out under rather different conditions, these temperatures were 1,360, 2,205, and 1,185°F, respectively. These examples also demonstrate how conservative estimates rather than "best estimates" are generally used in safety analyses. This is good engineering practice, but it is not usually recognized by those who use such estimates to frighten the public.

One type of LOCA in which the ECCS would not prevent a meltdown is a large crack in the bottom of the reactor vessel, since water injected by the ECCS would simply pour out through that crack. This would not occur with pipe breaks since all significant pipes enter the vessel near its top. This problem was intensively investigated by the British as part of their decision to convert from their own type to American-type reactors, and they concluded[15] that, in view of the large thicknesses (see Fig. 2) and high quality of the materials used, the probability of a large crack in the reactor vessel is so small as to be negligible. There is also an elaborate inspection program to ensure that the high quality of the reactor vessel material is maintained. One potential problem in this regard, "pressurized thermal shock," has received widespread publicity. It is discussed later in this chapter.

While every effort is being made to block the roads to meltdown, there is always a possibility of a road being opened by successive failures in the various lines of defense we have described. Or perhaps there is some obscure road to meltdown that no one has ever thought of in spite of the many years of technical effort on this problem. If nuclear power becomes a flourishing industry, there probably will be meltdowns somewhere someday. But if and when they occur, there is still one final line of defense—the containment—which should protect the public from harm in most cases. Let's now consider the reliability of that line of defense.

HOW SECURE IS THE CONTAINMENT?

In all of the accident scenarios we have considered, water and steam from inside the reactor pours out into the containment building. When water is pumped in by the emergency core cooling system, some of it overflows, and when it surrounds the fuel it boils into steam, which goes out through the break into the containment. We thus expect the containment to be filled with steam, with a lot of excess water on the floor. This is true in nearly all potential loss-of-coolant accidents, even if the system does not break open, as was the case in the Three Mile Island accident. In addition, heat is being fed into this water and steam by the radioactivity in the fuel, by chemical reactions of steam with the fuel casing, and by burning of the hydrogen generated in those reactions. The most important threat to the security of the containment is that this heat will raise the pressure of the steam to the point where it will exceed the holding power of the containment walls, about 10 times normal atmospheric pressure.

In order to counteract this threat, there are systems for cooling the containment atmosphere.[4,16] One such system sprays cool water into the air, a very efficient way of condensing steam; when it exhausts its stored water supply, it picks up water from the containment floor, cools it, and then sprays it into the air inside the containment. Another type of system consists of fans blowing containment air over tubes through which cool water is circulating. There are typically five of these systems, but only one (or in rare cases, two) need be operable in order to assure that the containment is adequately cooled. In most cases, one of the systems is driven by a diesel engine so as to be available in the event of an electric power failure. A more quantitative treatment of the containment cooling problem is given in the Chapter 6 Appendix.

Since they are safety related, these systems are subject to elaborate quality control during their manufacture and are frequently inspected and tested, so it seems reasonable to expect at least most of these systems to function properly if an accident should occur. All of them were functional during the Three Mile Island accident, and that is why it has been concluded[2] that the containment would have prevented the escape of radioactivity even if there had been a meltdown there.

Unfortunately, containment security is not always that favorable. Some of the accident scenarios outlined above also affect the containment heat removal systems. For example, electric power failure would prevent pumps from operating. But more important are accidents in which the containment is bypassed.[17] For example, in Fig. 3 we see that the steam generator contains tubes that are directly connected to the reactor, but the steam it generates passes out of the containment to the turbine. If the tubes rupture, there is then a direct path for the radioactivity in the reactor to get into the pipe leading to the turbine. If the reactor is still at its very high operating pressure, this high pressure could be transmitted into that pipe and break it open, releasing radioactivity outside the containment.

Two other mechanisms for breaking open the containment have been discussed. One of these is a steam explosion, which has received considerable research attention[7] and was publicized in the fictional movie "The China Syndrome." The worst situation is in a meltdown where the molten fuel falls into a pool of water at the bottom of the reactor vessel, producing so much steam so suddenly that the top of the reactor vessel would be blown off and hurled upward with so much force that it would break open the top of the containment building. This is a highly unlikely scenario.

In "The China Syndrome" it is implied that a sufficiently powerful steam explosion can occur when the molten fuel melts its way into the ground and comes into contact with groundwater. Actually, only a tiny fraction of the molten fuel would be coming into first contact with groundwater at any given time, so steam would be produced gradually rather than as a single explosive release. A fictional movie need not be realistic, of course, but it is important for the audience to recognize that point.

The movie also makes an issue of groundwater contamination following a meltdown accident. In actuality, were molten fuel to suddenly come into contact with groundwater, the latter would flash into steam, which would build up a pressure to keep the rest of the groundwater away. There would thus be little contact until the molten fuel cooled and solidified many days later. It would then be in the form of a glassy mass that would be highly insoluble in water, so there would be relatively little groundwater contamination. If that were judged to be a problem, there would be plenty of time to construct barriers to permanently isolate the radioactivity from groundwater thereafter. It is difficult to imagine a situation in which there would be any adverse health effects from groundwater contamination.

The other possible mechanism for breaking the containment, a hydrogen explosion, has received substantial research attention[18] and achieved notoriety in the Three Mile Island accident. The consensus of the research seems to be that even if all the hydrogen that could be generated in an accident were to explode at once, the forces would not be powerful enough to break most containments, including the one at Three Mile Island.[18] Moreover, in nearly all circumstances the hydrogen would be produced gradually, and there are many sources of sparks (e.g., electric motors) which would cause it to burn in a series of fires and/or small explosions not nearly large enough to threaten the containment.

Three of the U.S. pressurized water reactor (PWR) containments store large volumes of ice inside to reduce steam pressure in an accident. Since the presence of ice is a failproof method for cooling the surroundings and thereby avoiding high steam pressure, it was not considered necessary to build the containment walls so powerfully or to make the containment volumes so large. These containments are more vulnerable to a hydrogen explosion,[18] and therefore they are fitted with numerous gadgets for generating sparks to be extra certain that hydrogen ignites before large quantities can accumulate.[19]

The boiling water reactor (BWR) containments are much smaller in volume than those of PWRs; hence, they are more vulnerable to pressures generated by a hydrogen explosion. Some of these BWRs are operated with an inert gas filling the containment.[20] Since there is no oxygen present, hydrogen cannot combine with oxygen to explode. Other BWRs, however, are at substantial risk of containment failure due to hydrogen explosions, although the failure mode is such that most of the radioactivity would not escape.[17] It would exit through a pool of water which would dissolve and thus retain the radioactive materials.

Of course, explosions inside the containment, even if they do not crack the walls, can damage equipment, and this can cause problems. For example, if explosions disabled all of the heat removal systems, the containment might be broken by steam pressure. However, the probability for disabling many separate systems would be very small.

The systems we have described in this section and the previous one for averting a catastrophic accident constitute a "defense in depth," which is the guiding principle in designs for reactor safety. If the quality assurance fails, the inspections ordinarily provide safety. If the inspection programs fail, the leak detection saves the day. If that fails, the ECCS protects the system. And if the ECCS fails, the containment averts damage to the public. Moreover, each of these systems is itself a defense in depth; for example, if one of the ECCS water injection systems fails, another can do its job, and if both fail a third can provide sufficient water.

One sometimes hears statements to the effect that reactors are safe if everything goes right, but if any piece of equipment fails or if an operator makes a mistake, disaster will result. This statement is completely WRONG. In reactor design it is assumed that all sorts of things will go wrong — pipes will break, valves will stick, motors will fail, operators will push the wrong button, and so on, but there is "defense in depth" to cover these malfunctions or series of successive malfunctions.

Of course, the depth of the defense is not infinite. If each line of defense would crumble, one after the other, there could be a disaster. But as the depth of the defense is increased, the probability for this to happen is rapidly decreased. For example, if each line of defense has a chance of failure equal to that of drawing the ace of spades out of a deck of shuffled cards—one chance in 52, the probability for five successive lines of defense to fail is like the chance of drawing the ace of spades successively out of five decks of well-shuffled cards—one chance in $52 \times 52 \times 52 \times 52 \times 52$, or one chance in 380 million!

There have been cases where one of the lines of defense has failed in nuclear power plants. Utilities have been heavily fined by the NRC for such things as leaving a valve closed and thereby compromising the effectiveness of one of the emergency systems. These incidents are often given publicity as failures that could lead to a meltdown. But the media coverage rarely bothers to point out that there are several lines of defense remaining un-breached between these events and a meltdown—not to mention that there is still a major line of defense, the containment, remaining even if a meltdown occurs.

THE PROBABILITIES

In considering the hazards of a reactor meltdown accident, once again we find ourselves involved in a game of chance governed by the laws of probability. By setting up additional lines of defense, or by improving the ones we now have, we can reduce the probability of a major accident, but we can never reduce it to zero. This should not necessarily be discomforting since we already are engaged in innumerable other games of chance with disastrous consequences if we lose—natural phenomena like earthquakes and disease epidemics, and manmade threats like toxic chemical releases and dam failures, to name a few. In fact, participating in this new game of chance may save us from participating in others brought on by alternative actions, and it may therefore reduce our total risk: building a nuclear power plant may remove the need for a hydroelectric dam whose failure can cause a disaster, or for a coal-burning power plant whose air pollution might be disastrous. The important question is: what is the probability of a disastrous meltdown accident?

Several studies have been undertaken to answer this question. The best known of these was sponsored by the NRC and directed by Dr. Norman

Rasmussen, an MIT professor.[7] It extended over several years, involved many dozens of scientists and engineers, and cost over $4 million before its final report was issued in 1975. The report bore document designation "WASH-1400" and was titled "Reactor Safety Study" (RSS). It was a probabilistic risk analysis (PRA) based on a method known as "fault tree analysis," which had been developed to evaluate safety problems in the aerospace industry. It is described briefly in the Chapter 6 Appendix.

The history of the RSS does not stop in 1975. The Union of Concerned Scientists (UCS) published a critique of it[21] in 1977 with its own probabilities, and we will quote some of its conclusions. An independent review panel chaired by Professor Harold Lewis of the University of California was commissioned by the NRC and reported[22] in 1978. The principal finding of the Lewis panel was that the uncertainties in the probabilities given by the RSS were larger than originally stated, but that there is no reason to believe that the probabilities were either too large or too small. The Lewis panel also took exception to the 12-page Executive Summary issued with the RSS. The NRC accepted the Lewis panel report in 1979, so in our references to the RSS we will not use either the Executive Summary or the uncertainty estimates. In the late 1970s there were similar RSSs carried out in West Germany and in Sweden, using similar methodology and obtaining similar results. During this period, the RSS was pooh-poohed by opponents of nuclear power and was interpreted by the media and hence by the public as controversial.

Nevertheless, the NRC continued to encourage development and improvement of PRA methodology, and PRAs were carried out for over a dozen U.S. reactors at a typical cost of $5 million each (the RSS had analyzed only two reactors, one PWR and one BWR). With all of this activity and effort, new ideas and procedures were developed, and older ones were shown to be wanting and were abandoned. During this period important new scientific and technical developments surfaced, and these were incorporated. For example, there were intensive studies of the chemical form of released radioactivity,[23] which had only been guessed at in the RSS, and greatly improved estimates were obtained on the strength of containments. A new appreciation of the importance of earthquakes and fires arose,[17] and extensive research was devoted to analysis of their impact.

While the RSS was carried out by one team in 2 years at a time when PRA methodology was in its infancy and when there had been very little operating experience with nuclear reactors, these newer studies were done by many different teams over periods of more like 10 years, when PRA methodology was much more mature and there was many times as much operational experience, including several "accidents" of various magnitude, by far the worst of which was the one at Three Mile Island.

One interesting new development has been abandonment of the word *meltdown*, largely replaced by *core damage*. In the early thinking about reactor accidents, the idea became prevalent that if any appreciable fuel melting would occur, the problem would continue to escalate until all of the fuel became a molten mass with an unstoppable internal heat source (the radioactivity). Hence it would melt its way through the reactor vessel and anything else that got in its way—down through the Earth and all the way to China was the picturesque exaggeration that led to the name "China Syndrome." More detailed studies showed that these ideas were grossly oversimplified, and the Three Mile Island accident was a clear counterexample—most of the fuel melted, but it did not even get out of the reactor vessel. It is even difficult to answer the question "Was the Three Mile Island accident a meltdown?" because that word is not clearly defined. "Core damage," on the other hand, allows discussion of the wide variety of circumstances that are now believed to be possible. It also allows consideration of the several "precursors" to core damage that have already been experienced in reactor operation. By noting what further failures could have caused these incidents to escalate into core damage and estimating the probabilities for these further failures, one can arrive at an independent estimate of the probability for a core damage accident. The results of the new PRAs are discussed in some detail in the Chapter 6 Appendix. There are many differences between these and the RSS, but when all is said and done, the bottom lines turn out to be quite similar.[17] It is therefore not unreasonable to use the RSS results. There is a big advantage in doing so since the RSS gives many more details that are useful in the discussion. We therefore base the following discussion on the RSS.

The RSS estimates that a reactor meltdown may be expected about once every 20,000 years of reactor operation; that is, if there were 100 reactors, there would be a meltdown once in 200 years.[7] The report by the principal organization opposed to nuclear power, Union of Concerned Scientists (UCS),[21] estimates one meltdown for every 2,000 years of reactor operation. In U.S.-type reactors, there have been over 2,000 years of commercial reactor operation worldwide plus almost 4,000 years of U.S. Navy reactor operation, all without a meltdown (in the sense they are using the word). If the UCS estimate is correct, we should have expected three meltdowns by now, whereas according to the RSS, there is a 30% chance that we would have had one.

We now turn to the consequences of a meltdown. Since it gives more detail, we will quote the results of the RSS here; the UCS viewpoint can be roughly interpreted as multiplying all consequences by a factor of 10.

In most meltdowns the containment is expected to maintain its integrity for a long time, so the number of fatalities should be zero. In 1 out of 5 meltdowns there would be over 1,000 deaths, in 1 out of 100 there would be over 10,000 deaths, and in 1 out of 100,000 meltdowns, we would approach 50,000 deaths (the number we get each year from motor vehicle accidents). Considering all types, we expect an average of 400 fatalities per meltdown; the UCS estimate is 5,000. Since air pollution from coal burning is estimated to be causing 30,000 deaths each year in the United States (see Chapter 3), for nuclear power to be as dangerous as coal burning there would have to be 75 meltdowns per year ($30,000 \div 400 = 75$), or 1 meltdown every 5 days somewhere in the United States, according to the RSS; according to UCS, there would have to be a meltdown every 2 months. Since there has never been a single meltdown, clearly we cannot expect one nearly that often.

It is often argued that the deaths from air pollution are not very alarming because they are not detectable, and we cannot associate any particular deaths with coal burning. But the same is true of the vast majority of deaths from nuclear reactor accidents. They would materialize only as slight increases of the cancer rate in a large population. Even in the worst accident considered in the RSS, expected only once in 100,000 meltdowns, the 45,000 cancer deaths would occur among a population of about 10 million, with each individual's risk being increased by 0.5%. Typically, this would increase a person's risk of dying from cancer from 20.0% to 20.5%. This risk varies much more than that from state to state — 17.5% in Colorado and New Mexico, 19% in Kentucky, Tennessee, and Texas, 22% in New York, and 24% in Connecticut and Rhode Island — and these variations are rarely, if ever, noticed. It is thus reasonable to assume that the additional cancer risks, even to those involved in this most serious meltdown accident considered in the RSS, would never be noticed.

If we are interested in detectable deaths that can be attributed to an accident, we must limit our consideration to acute radiation sickness, which can be induced by very high radiation doses, about a half million millirems in one day resulting in death within a month. This is a rather rare disease: there were three deaths due to it in the early years among workers in U.S. government nuclear programs, but there have been none for over 25 years now.

According to the RSS, there would be no detectable deaths in 98 out of 100 meltdowns, there would be over 100 such deaths in one out of 500 meltdowns, over 1,000 in one out of 5,000 meltdowns, and in one out of 100,000 meltdowns there would be about 3,500 detectable fatalities.

The largest number of detectable fatalities to date from an energy-related incident was an air pollution episode in London in 1952 in which 3,500 deaths directly attributable to the pollution occurred within a few days.[24] Thus, with regard to detectable fatalities, the equivalent of the worst nuclear accident considered in the RSS—expected once in 100,000 melt-downs—has already occurred with coal burning.

But the nuclear accidents we have been discussing are hypothetical, and if we want to consider hypothetical accidents, very high consequences are not difficult to find. For example there are at least two hydroelectric dams in the United States whose sudden rupture would kill over 200,000 people.[7] There are hypothetical explosions of liquefied natural gas that can wipe out a whole city. If we get into possibilities of incubating or spreading germs, or of subtle chemical effects, we can easily imagine even more devastating scenarios arising due to air pollution from coal or oil burning plants.

It is sometimes said that nuclear accidents may be extremely rare, but when they occur they are so devastating as to make the whole technology unacceptable. From the above comparisons it is clear that this argument holds no water. For another perspective, we embrace a technology that kills 50,000 Americans every year. Every one of these deaths is clearly detectable, and that technology seriously injures more than 10 times that many. I refer here to motor vehicles. Even if we had a meltdown every 10 years, a nuclear power accident would kill that many only once in a million years.

THE WORST POSSIBLE ACCIDENT

One subject we have not discussed here is the "worst possible nuclear accident," because there is no such thing. In any field of endeavor, it is easy to concoct a possible accident scenario that is worse than anything that has been previously proposed, although it will be of lower probability. One can imagine a gasoline spill causing a fire that would wipe out a whole city, killing most of its inhabitants. It might require a lot of improbable circum-stances combining together, like water lines being frozen to prevent effec-tive fire fighting, a traffic jam aggravated by street construction or traffic accidents limiting access to fire fighters, some substandard gas lines which the heat from the fire caused to leak, a high wind frequently shifting to spread the fire in all directions, a strong atmospheric temperature inversion after the whole city has become engulfed in flame to keep the smoke close to the ground, a lot of bridges and tunnels closed for various reasons, eliminating escape routes, some errors in advising the public, and so forth. Each of these situations is improbable, so a combination of many of them occurring in sequence is highly improbable, but it is certainly not impossible.

If anyone thinks that is the worst possible consequence of a gasoline spill, consider the possibility of the fire being spread by glowing embers to other cities which were left without protection because their firefighters were off assisting the first city; or of a disease epidemic spawned by unsanitary conditions left by the conflagration spreading over the country; or of communications foul-ups and misunderstandings caused by the fire leading to an exchange of nuclear weapon strikes. There is virtually no limit to the damage that is possible from a gasoline spill. But as the damage envisioned increases, the number of improbable circumstances required increases, so the probability for the eventuality becomes smaller and smaller. There is no such thing as the "worst possible accident," and any consideration of what terrible accidents are possible without simultaneously considering their low probability is a ridiculous exercise that can lead to completely deceptive conclusions.

The same reasoning applies to nuclear reactor accidents. Situations causing any number of deaths are possible, but the greater the consequences, the lower is the probability. The worst accident the RSS considered would cause about 50,000 deaths, with a probability of one occurrence in a billion years of reactor operation. A person's risk of being a victim of such an accident is 20,000 times less than the risk of being killed by lightning, and 1,000 times less than the risk of death from an airplane crashing into his or her house.[7]

But this once-in-a-billion-year accident is practically the only nuclear reactor accident ever discussed in the media. When it is discussed, its probability is hardly ever mentioned, and many people, including Helen Caldicott, who wrote a book on the subject, imply that it's the consequence of an average meltdown rather than of 1 out of 100,000 meltdowns. I have frequently been told that the probability doesn't matter—the very fact that such an accident is possible makes nuclear power unacceptable. According to that way of thinking, we have shown that the use of gasoline is not acceptable, and almost any human activity can similarly be shown to be unacceptable. If probability didn't matter, we would all die tomorrow from any one of thousands of dangers we live with constantly.

LAND CONTAMINATION

Another aspect of a reactor meltdown accident that has been widely publicized is land contamination. The most common media version is that it would contaminate an area the size of the state of Pennsylvania, 45,000 square miles. Of course this depends on one's definition of "contaminate." It could be said that the whole world is contaminated, because there is natural radioactivity everywhere; or that the state of Colorado is contaminated because the natural radiation there is twice as high as in most other states. However, the Federal Radiation Council in the United States and similar official agencies in other countries have adopted criteria for the upper level of contamination that is acceptable before people must be evacuated. This level corresponds roughly to doubling or tripling the average lifetime dose that would be received from natural radiation and medical X-rays, or 2 to 5 times as much extra radiation as would be received by the average American from moving to Colorado. It is still 4 to 10 times less than the natural radiation received by people living in some areas of India and Brazil. Studies of these people have given no evidence of health problems from their radiation exposure.[25]

With this definition, the worst meltdown accident considered[7] in the RSS—about 1% of all meltdowns might be this bad—would contaminate an area of 3,000 square miles, the area of a circle with a 30-mile radius. About 90% of this area could be cleaned up by simply using fire hoses on built-up areas, and plowing the open ground, but people would have to be relocated from the remaining 10%, an area equal to that of a circle with a 10-mile radius.

In assessing the impacts of this land contamination, I believe the appropriate measure is the monetary cost; the cost of decontaminating, relocating people, compensating for lost property and lost working time, buying up and destroying contaminated farm products, and so on. Some might argue that it is unfair to concentrate on money and ignore the human problems in relocation, but that is part of reality. Forced relocation is a common practice in building hydroelectric dams (which flood large land areas), highway construction, slum clearance projects, and so forth, and in these contexts the monetary cost and advantages to be gained are always the prime consideration in deciding on whether to undertake the project.

In most meltdowns, the cost would be less than $50 million (all costs are in 1975 dollars); in 1 out of 10 meltdowns, it would exceed $300 million; in 1 out of 100 meltdowns, it would exceed $2 billion; and once in 10,000 meltdowns, it would be as much as $15 billion.

Over all cases, the average cost would be about $100 million. Generating electricity by coal burning is estimated[26] to do about $600 million per year in property damage, destroying clothing, eroding building materials, and so forth. Thus it would require six meltdowns per year—one every 2 months—for the monetary cost to the public from reactor accidents to equal that from coal burning. Clearly, health impacts are more important than property damage in determining the risks of generating electricity, but the relative risks of nuclear power and coal are not very different for the two.

PUBLIC MISUNDERSTANDING

In this chapter we have shown that there have been serious misunderstandings of reactor meltdown accidents in the public mind. In most such accidents there would be no harm to the public, and the average meltdown would cause 400 fatalities and do $100 million in off-site damage. Even in the worst 0.001% of accidents, the increased cancer risk to those involved is much less than that of moving from other parts of the country to New England. This is a far cry from the public image of many thousands of dead bodies lying around in a vast area of devastation, and it certainly is not "the ultimate disaster." Only a tiny fraction of the public recognizes that for nuclear accidents to be as dangerous as coal burning, we would have to experience a meltdown every 5 days.

The consequences of the misunderstandings have been tragic. Surely no one believes that we will have a meltdown every 5 days, or even every few months. We have never even had a large scale evacuation, which would be the first step if there was any apparent danger to the public. Mass evacuations following other types of accidents are quite common. Chemical spills lead to evacuation of hundreds of people several times per year in the United States. In 1979, as a result of an accident of a railroad tank car carrying a dangerous chemical, there was a mass evacuation from a suburb of Toronto, involving over 100,000 people for several days.

Nevertheless, because of the misunderstandings attending nuclear accidents, utilities have continued to build coal-fired rather than nuclear plants. Every time this is done, thousands of Americans are condemned to a premature death.

NONSAFETY ISSUES

Any new technology is bound to encounter numerous technical problems that must be ironed out, and there has never been any reason to believe that nuclear technology should be an exception in this regard. However, contrary to the situation in other industries, technical problems in the nuclear industry often received widespread media exposure, causing them to be interpreted as safety issues.

Nearly any technical problem can indeed become a safety issue if it is consistently ignored. If an automobile runs out of lubricating oil, it could stall on a railroad crossing, which is clearly a safety problem. But the oil level is easy to check, there is a warning light indicating loss of oil pressure, and if the oil did run out, ominous grinding noises would alert you before the car would stall. Loss of lube oil is therefore not ordinarily considered to be a safety problem. It can be inconvenient, costly to fix, and may cause expensive damage to the engine, but it surely ranks far down on any list of safety hazards in automobiles. However, if the problem were not so familiar to a large segment of the population, the publicizing of one such case could easily scare people with stories about the possibility of automobiles stalling on railroad crossings or in other precarious situations due to loss of lube oil.

Analogous situations have been reported as safety issues for the nuclear industry. Let us review a few of them here.

Pressurized Thermal Shock[27]

The thick steel vessel housing the reactor is normally very hot because of the high temperature of the water inside (600°F). If, due to some malfunction, the inside is suddenly filled with cool water, the vessel experiences what is called "thermal shock." If it is then subjected to high pressure—producing pressurized thermal shock (PTS)—there is an increased tendency for the vessel to crack, rather than simply to stretch, if a small crack or imperfection already exists. The importance of PTS problems depends on quantitative details—how much of a thermal shock followed by how much pressure causes how much of an increased tendency to crack. Under ordinary conditions these quantitative details indicate that there is nothing to be concerned about. However, just as radiation can damage biological tissue, it can damage steel by knocking electrons and atoms out of their normal locations. This radiation damage to the reactor vessel aggravates its susceptibility to PTS.

Scientists recognized this problem over 20 years ago and they found a simple remedy for it—reducing the quantity of copper in the steel alloy from which the vessel is fabricated. This remedy was implemented in 1971, and all reactor vessels fabricated since that time have had no problems with PTS.

Reactor vessels fabricated before 1971 are kept under periodic observation to keep track of the problem. For many years, the NRC, burdened by other more urgent problems, put off considering PTS by adopting a very conservative screening criterion to indicate when further action on it would be undertaken. In 1981, time for action according to that criterion was only 1 or 2 years away in some reactors; hence, the NRC began to look into the problem in more detail by requesting information from various power plants. Misinterpreting these requests, a prominent newspaper ran a page-one story[28] headlined "Steel Turned Brittle by Radiation Called a Peril at 13 Nuclear Plants," broadly implying that serious safety problems were immediately at issue. Opponents of nuclear power soon began trumpeting that message. They claimed that reactor vessels would crumble like glass under PTS, although no such behavior has ever been observed in the numerous laboratory tests of PTS. In 1981-1982, the NRC and the nuclear industry delved into the PTS problem rather deeply. In 1982, the NRC came up with new conclusions and regulations.

When the radiation damage reaches the stage where action is required, several remedies are available, although not all are applicable in all situations. One way to postpone the problem is to redistribute the fuel in the reactor so as to reduce the radiation striking the walls of the vessel—this is now being done in several plants. One remedy for PTS is to keep the water storage tanks heated to reduce the thermal shock that would be caused by sudden water injections. Another option is to change operating procedures to reduce the suddenness with which this water can be introduced. The most complete remedy, which is also the most time consuming and expensive, is to heat the reactor vessel to a very high temperature (850°F) to anneal out the radiation damage; this would, in fact, make the vessel as good as new.

The NRC standard is a conservative one. It is based on the assumption that there is a small crack or flaw in the vessel, although these vessels are very carefully inspected and no small cracks or flaws have been found. The vessel is typically 8 inches thick, so the outside is exposed to considerably less radiation and thermal shock than the inside; therefore, even if there should be cracking inside, it would probably not extend all the way through the thickness of the vessel and there would consequently be no danger from it.

As long as the problem is recognized, is under constant surveillance, has remedies, and will not be allowed to reach the danger point, it seems fair to classify pressurized thermal shock as a technical problem rather than as a safety issue. It should therefore receive the attention of scientists and engineers, but there is no reason for the public to preoccupy itself with it.

Stress Corrosion Cracking of Pipes[29]

There have been a number of situations in which pipes in boiling water reactors have been found to have cracks. Since a pipe cracking open is a widely heralded potential cause for a LOCA, this problem has received extensive media coverage as a potential threat, especially when the first such crack was discovered in 1975. However, researchers have established that this type of cracking develops very slowly and is easily detected by ultrasonic tests in its very initial stages. If not, it leads to slow leaks which are readily detected and repaired. Stress corrosion cracking is therefore not a safety issue.

On the other hand, this problem has caused expensive shutdowns for repairs, and has therefore been an important problem for power plant owners. They have consequently invested tens of millions of dollars on research to overcome it. The first fruit of this research was to gain an understanding of the problem: welding stainless steel pipe joints caused some of the chromium that makes that material corrosion resistant to migrate away, reducing its local concentration from the normal 17% to below the 12% minimum for resistance to corrosion by excess oxygen in the water. Moreover, once this migration of chromium is started by the welding, the heat of the reactor water continues the process. A combination of this corrosion with stress on the material was found to cause the cracking.

Once the problem was understood, researchers rapidly found solutions. A new alloy with less carbon and more nitrogen, called nuclear-grade stainless steel, was developed which virtually eliminates the problem in new pipe. Investigators found that in the old type pipe, the chromium migration could be reversed by heating the welded joint in a furnace to 1,950°F, or by putting a lining of weld metal inside the pipe before the outside is welded. In addition to avoiding the chromium migration, methods have been developed to relieve the stress by running cooling water inside the pipe while the joint is being welded, or by heating the outside of the pipe while cooling the inside after the welding is completed. This last method is applicable without removing installed pipes. All of these methods are now being applied in operating plants. Moreover, researchers are developing methods for reducing the free oxygen content in the water, the principal chemical agent responsible for the corrosion. All three factors, chromium migration, mechanical stress, and a corrosive chemical agent, are necessary to cause the cracking, and all three of them have been reduced by these measures. An automated ultrasonic testing system has been developed to predict which welds are most likely to fail and to estimate their remaining service life. All this progress has put stress corrosion cracking of pipes well under control.

Steam Generator Tube Leaks[30]

A diagram of a pressurized water reactor (PWR) is shown in Fig. 3. The water in the reactor is kept under sufficiently high pressure that it does not boil and become steam. Rather it is pumped through the tubes of "steam generators" where it transfers its heat to the water from a separate "secondary" system, causing the latter to boil into steam. This has some advantage (and some disadvantages) over the simpler system of generating the steam by boiling the water in the reactor, as in the BWR. One of the advantages is that the water from the reactor, which contains radioactive contaminants, never gets into the other areas of the plant (turbine, condenser, etc.), so less attention to radioactivity control is needed in those areas.

However, leaks in steam generator tubes do allow radioactivity to reach those areas, and since they have minimal radioactivity control, it can easily escape from there into the environment. A large fraction of American PWRs have experienced problems with steam generator tube leaks. There are many thousands of these tubes in a steam generator; therefore, leaking tubes can simply be plugged-up at both ends without affecting operation. However, when the number of plugged tubes exceeds about 20% of the total, as it has in some plants, the electrical generating capacity is significantly reduced. This represents a costly loss of revenue to the utility. In at least three cases, the utility has decided to replace their steam generators, a rather expensive alternative requiring many months of shutdown.

From the safety viewpoint, the worst accident worthy of consideration in this area is a sudden complete rupture of a few tubes. Such an accident might be expected once every several years. This is what happened at the Ginna plant near Rochester, New York, in January 1982. That accident generated a great deal of publicity, but the maximum exposure at any off-site point was 0.5 mrem,[31] less than the average American receives from natural sources every day. Since there were no people staying all day at such points, no member of the public received even that much exposure. The total of the exposures to the whole population in the area was less than 100 mrem, which gives only 1 chance in 80,000 that there will ever be a single death resulting. On the other hand, it has been a costly problem for utilities, and a great deal of research has been devoted to solving it.

Eight separate classes of failures have been identified—denting, erosion-corrosion, fatigue, fretting, intergranular attack, pitting, stress corrosion cracking, and wastage. Researchers have developed a number of different methods for reducing these problems and for avoiding them in new plants. They have also developed new methods for detecting, locating, evaluating, and repairing leaks.

The NRC keeps a close watch on these problems to be certain that public safety is not compromised, in spite of the very small potential of steam generator leaks to cause radiation exposure to the public. It requires frequent testing for leaks, and has strict limits on the amount of leakage that can be tolerated before the reactor is shut down for repairs. It also maintains research programs to achieve improved understanding, evaluations, and predictability of future problems. The industry itself is also doing a great deal of research on the problem.

Chapter 7 / THE CHERNOBYL ACCIDENT—CAN IT HAPPEN HERE?

It is very difficult to predict the future of scientific developments, and few would even dare to make predictions extending beyond the next 50 years. However, based on everything we know now, one can make a strong case for the thesis that nuclear fission reactors will be providing a large fraction of our energy needs for the next million years. If that should come to pass, a history of energy production written at that remote date may well record that the worst reactor accident of all time occurred at Chernobyl, USSR, in April of 1986.

In that accident, a substantial fraction of all of the radioactivity in the reactor was dispersed into the environment as airborne dust—its most dangerous form. It is difficult to imagine how anything worse could happen to a reactor from the standpoint of harming the public outside.

In the wake of the Chernobyl accident, the primary question on American minds was—can it happen here? Let us try to answer that question.

We have just seen how extremely improbable an accident of that magnitude should be. But if it is so extremely improbable, how could it have happened so early in the history of nuclear power? The response to that question is that there are very major differences between the Chernobyl reactor and the American reactors on which our previous discussion was based.

In order to understand these differences, we must delve much deeper into the details of how reactors work. This discussion may also be useful to those with an interest in the basic science behind nuclear power.

HOW NUCLEAR REACTORS WORK

In an ordinary furnace, energy is produced in the form of heat by chemical reactions between the fuel and oxygen in the air. A chemical reaction is actually a collision between atoms in which their orbiting electrons interact. The other constituent of an atom is the nucleus. If two nuclei collide and interact we have a *nuclear* reaction. However, unlike atoms, which are electrically neutral, nuclei have a positive electric charge and therefore strongly repel one another. Hence nuclear reactions do not normally occur in our familiar world.

An exception to this situation is the neutron, one of the two constituents of nuclei (the other is the proton), which does not have an electric charge. It can therefore approach a nucleus without being repelled and induce a nuclear reaction. Because this happens so easily, a neutron can move about freely for only about 0.0001 seconds before it collides with a nucleus and becomes involved in a nuclear reaction. Since free neutrons last for such a short time, they must be produced as they are used. Neutrons can only be produced in nuclear reactions, so what is needed is a nuclear reaction induced by a neutron which releases more than one neutron. These can then induce further reactions which produce more neutrons, and so forth, in a self-sustaining chain reaction. Such a reaction is available in the interaction of a neutron with a uranium-235 (U-235) nucleus. This is the basis for a nuclear reactor.

When a U-235 nucleus is struck by a neutron, it often splits into two nuclei of roughly half the size and mass in a process called "fission." Since all nuclei have a positive electrical charge, these two newly formed nuclei repel one another very strongly. As a result they end up travelling in opposite directions at very high speeds, which means that their motion contains lots of energy. As they travel through the surrounding material, whatever it may be, they strike other atoms, giving them some of their energy, until, after about a million such collisions over a few thousandths of an inch of travel, all of their energy is dissipated, and they come to rest. The atoms they strike or their orbiting electrons are given additional motion and have collisions with other atoms, sharing their energy with them. By these processes, the energy released in the fission process is eventually shared by all of the atoms in the vicinity. It increases the speed of their normal random motion and our senses interpret this as increased temperature. Thus, the fission reaction releases heat, 50 million times as much heat as is released in the chemical reaction between a carbon atom from coal and oxygen atoms from the air in the coal-burning process. The purpose of a nuclear power plant is to convert this heat into electricity, as we described in Chapter 6.

The two original fragments from the fission process also have a substantial excess of *internal* energy which they largely dissipate by shooting off neutrons, typically two or three neutrons from each fission reaction. It is these neutrons that sustain the chain reaction. In order for it to be self-sustaining, at least one of them must strike another U-235 nucleus and cause a fission reaction. Some neutrons get past the surrounding U-235 and are lost to the process. If enough neutrons are lost, the chain reaction will stop. These losses are reduced as the thickness of the U-235 that the neutron must traverse increases. This means that for the chain reaction to be self-sustaining, there must by some minimum amount of U-235. This is called the *critical mass*. To generate energy, one need only assemble a critical mass of U-235, which is about the size of a cantaloupe, and introduce a few neutrons to start the process. There are simple and readily available ways of providing these start-up neutrons.

But where do we get the U-235? Uranium occurs in nature as a mixture of 99.3% uranium-238 (U-238) with 0.7% U-235. When a neutron strikes U-238, that nucleus does *not* undergo fission. If we assemble a large mass of natural uranium, we do not get a self-sustaining chain reaction because the great majority of neutrons are lost by striking U-238 nuclei. As one possible solution to this problem we can separate the U-235 out of natural uranium; we do this for making bombs, but it is a very difficult and expensive process.

However, an alternative and much better approach is available. If the neutrons can be slowed down to very low speeds—one ten-thousandth of the velocity with which they originally emerge—due to the quirks of quantum physics, their inherent probability for striking a U-235 nucleus becomes 200 times greater than for striking a U-238 nucleus. In this situation, even with natural uranium most neutrons would strike U-235 nuclei, and we could get a chain reaction.

The method for slowing down neutrons is to arrange for them to strike and bounce off lightweight nuclei, giving the struck nuclei some of their energy. Materials introduced for this purpose are called "moderators" since they moderate the speed of the neutrons. When a neutron strikes any nucleus, there is some chance that it will be absorbed, but the probability varies by large factors for different nuclei. Since we cannot afford to lose many neutrons, a moderator is only suitable if it has a low probability for neutron absorption. This leaves very few options. One of these is very high purity carbon in the form of graphite. It is such a good moderator that natural uranium dispersed in very high purity graphite can provide a chain reaction. That is how the first chain reaction was achieved in the famous experiments directed by Enrico Fermi under the stands of the University of Chicago football stadium in 1942. (That reactor may be seen at the Smithsonian Museum in Washington.) Another possible candidate for a moderator is ordinary water, but its propensity for capturing neutrons is not as low as one would like. A chain reaction cannot be achieved from a mixture of natural uranium and water. (Actually, this is fortunate because if it could be achieved, reactors would be very easy to make and Hitler would have had nuclear bombs during World War II.) However, if the uranium is enriched in U-235 up to 3% (from its normal 0.7%), then water becomes a good moderator. It turns out that providing this relatively low enrichment is not prohibitively expensive.

U.S. REACTORS VERSUS CHERNOBYL-TYPE REACTORS[1]

From the foregoing discussion, we see that two of the principal options for reactor design are:

1. Uranium fuel enriched to 3% in U-235 surrounded by water moderator; this is the option used in all U.S. power reactors.
2. Natural (or slightly enriched) uranium surrounded by graphite moderator; this is the option used in Chernobyl-type reactors.

Since the heat is generated in the uranium fuel, there is still the problem of transferring this heat out of the reactor to make the steam which drives the turbine to produce electricity. This is done efficiently by circulating water as in the case of cooling an automobile engine, but on a much grander scale. Option 1 thus becomes a configuration of fuel rods in a large water-filled vessel with water being rapidly pumped through. That is what is done in all U.S. reactors. Option 2, which is used in Chernobyl-type reactors, consists of a large block of graphite with holes in it containing tubes; these tubes have fuel rods inside of them, and water flows rapidly through the tubes to remove the heat. This water provides no benefit as a moderator since the graphite takes care of that function. On the other hand, the water does capture neutrons, reducing the number of neutrons available for striking uranium atoms. The net effect of the water on the chain reaction is, therefore, negative, tending to slow it down. Materials that act in this way are called "poisons," since they tend to destroy, or poison, the chain reaction. In a Chernobyl-type reactor, the water acts as a "poison."

There are some important safety advantages to Option 1 which is the U.S. approach. If, due to an accident, the water should be lost, the chain reaction automatically stops—there can be no chain reaction without the moderator. However, in Option 2, the Chernobyl design, the graphite moderator is still there, and loss of water means loss of a "poison." Losing a poison speeds up the chain reaction. This generates additional heat at a time when the mechanism for removing the heat—the water—is gone. This can be a very dangerous situation.

Another safety advantage of the U.S. approach is that if, for any reason, the chain reaction speeds up, releasing more energy and thus causing the temperature to rise, the water acts as a buffer. The increased temperature will cause more boiling. This will reduce the amount of moderator, which will slow down the chain reaction and thereby reduce the temperature. The reactor is, therefore, *stable* against a temperature change; that is, an increase in temperature automatically causes things to happen which will reduce the temperature. No human action or equipment failure can interfere with this natural process.

In a Chernobyl-type reactor, on the other hand, an increase in the speed of the chain reaction causes the temperature to increase, which causes more water boiling. This reduces the amount of "poison," which causes the chain reaction to accelerate and increases the temperature even further. This process, therefore, tends to make the reactor *un*stable against a temperature change; an increase in temperature automatically causes things to happen which lead to further increases in the temperature. Something must be done by some person or equipment to prevent the situation from escalating to a disaster. Actually, under normal operating conditions, other factors would contribute to overcome this instability, but in low-power operation, where the infamous accident occurred, this instability represented an extremely dangerous safety problem.

With these two very clear safety advantages for the U.S.-type reactors, one might ask why anyone would build a Chernobyl-type reactor. The reason is that Chernobyl-type reactors are designed to produce plutonium for bombs while they generate electricity. This type of reactor has two big advantages for this application.[1] One is that the quantity of plutonium produced varies inversely with the ratio of U-235 to U-238, which means that much more plutonium is produced in Chernobyl-type reactors than in U.S. reactors. The other is that in producing plutonium for bombs, it is important that the fuel be left in the reactor no more than 30 days, and a Chernobyl-type reactor is much better adapted for that purpose.

In a U.S. reactor, all of the fuel is inside a single large vessel, and it is a major effort, requiring about a month's time, to shut down the reactor, open the vessel, and change the fuel. Therefore, this operation is undertaken no more than once a year, which makes these reactors unsuitable for producing weapons-grade plutonium. In a Chernobyl-type reactor, each of the 1,700 fuel rods is enclosed in a single tube through which the water flows. It is relatively easy to open one of these tubes at a time, change the fuel rod, and replace it, without having to shut down the reactor. This makes these reactors excellent facilities for producing bomb-grade plutonium as they generate electricity. In fact, some of the U.S. government reactors designed only to produce plutonium for bombs are somewhat like the Chernobyl-type reactor. After the Chernobyl accident, there were serious questions raised about safety hazards in these U.S. production reactors, but it was eventually concluded that they contain design features that assure their safety.

However, there is one further price in safety that must be paid for the capability to change fuel easily. The fuel-changing operation requires a lot of space and activity by operators. This makes it impractical to enclose the reactor in the type of containment used for U.S. reactors (as described in Chapter 6). The containment used in a Chernobyl-type reactor is designed only to protect against rupture of one of the 1,700 tubes, rather than against a major accident that may rupture hundreds of tubes. All of the added safety obtained from containments in U.S. reactors was, therefore, not available at Chernobyl. In fact, post-accident analyses indicate that if there had been a U.S.-style containment, none of the radioactivity would have escaped, and there would have been no injuries or deaths.

THE CHERNOBYL ACCIDENT — BLOW-BY-BLOW[2]

In April 1986, it was decided to use the Chernobyl power plant for an electrical engineering experiment on its turbine-generator, the machinery used to convert the energy of steam into electricity. The purpose was to develop a system for utilizing the rotational inertia of the turbine-generator to operate water pumps if electric power should be lost. The only function of the reactor was to get the rotation of the turbine and generator up to speed before beginning the experiment. Since no experimentation with the reactor was involved, no reactor experts were on hand. Electrical engineers supervised the experimental work while the reactor was run by the regular operators.

The experiment was set to start at 1:00 P.M. on April 25, but a need for the plant's electrical output developed unexpectedly, delaying the experiment until 11:00 P.M. At that time, the power level of the reactor was reduced to the level desired for the experiment, but in the operators' rush to make up for lost time, they reduced the power too rapidly.

Reactors have a peculiar characteristic: if they are shut down, a neutron-absorbing "poison" develops that prevents them from being restarted for many hours. The overly rapid reduction in power led to a build-up of this poison that made it difficult to get anywhere near the desired power level, 25% of full power. In order to get as much power out of the reactor as possible, many of the control rods had to be withdrawn, but still, the power level was only about 6% of full power (one-fourth of the power level planned for the experiment). At this low power level the temperature instability becomes very pronounced, and it was, therefore, strictly against the plant rules to operate under those conditions.

Nevertheless, at 1:00 A.M. (April 26) the supervisors decided to go ahead with the experiment. At 1:05 A.M., additional water pumps were turned on as part of the experiment; these were the pumps to be driven by the rotational inertia of the turbine-generator following a loss of electricity. No one seemed to notice that this action was providing too much water flow for the reactor at this lower power level. In fact, that quantity of water flow at such a low power level was forbidden by the rules. Coincidentally, a normal operating situation came up which, by a quirk in the reactor design, caused a further increase in water flow at 1:19 A.M. Since water acts as a poison, this additional water flow required withdrawal of the manual control rods. That put the reactor into a condition such that a loss of water would make it "prompt critical," which means that the power would escalate very rapidly, doubling every second or so. Operating any reactor in that condition is strictly prohibited, but apparently ignoring rules was not considered to be a serious transgression at the Chernobyl plant.

At 1:22, the added water flow started at 1:19 was stopped, but since it takes a minute or two for this to affect the conditions in the reactor, the manual control rods were not immediately re-inserted. The very dangerous operating condition continued. At 1:22:30, a computer printed out a warning that the reactor's condition was unsafe and that it should be shut down immediately, but the operator decided to ignore it, a very serious rule violation. It is difficult to understand why he ignored it, but we will never know because he died in the accident. At 1:23:04 the experiment began.

One of the effects of the experiment was to cause the water pumps turned on at 1:05 A.M. to slow down—their normal electrical drive was shut off, and they were then being driven by the rotational inertia of the turbine-generator. The slow-down of these pumps reduced the water flow to the reactor, which caused more of the water in the fuel tubes to turn into steam. Another effect of the experiment was to cut off the flow of steam to the turbine—it was to continue rotating only by inertia—which increases the amount of steam in the reactor. All reactors have an interlock which automatically shuts down the chain reaction when the steam supply to the turbine is cut off, but that interlock was disabled for the experiment (as were several others). At about this time, the reduction in water flow at 1:22 A.M. began to have an effect, which further increased the amount of steam in the reactor. Since water is a poison, converting water to steam reduces the amount of poison in the reactor, causing the chain reaction to speed up. In reaction to this, the automatic control rods went all the way in, which gives the maximum effect they can provide to reduce power, but the chain reaction still continued to speed up, accentuated by the instability against temperature increase described above. The increase in temperature caused more water to boil into steam, which further accelerated the chain reaction, and further increased the temperature.

If the manual control rods had not been withdrawn, there would have been no problem, but unfortunately there was not enough time to reinsert them. At 1:23:40, the order was given to insert the emergency shutdown control rods. However, because of a questionable aspect in their design, their insertion is rather slow, only one-fourth as fast as in U.S. reactors. Before they got all the way in, they were blocked by damage that had already occurred. Some of the tubes had burst. The speed of the chain reaction continued to escalate, and there was no way to stop it. In a U.S. reactor, loss of water would have meant loss of the moderator, which would have stopped the chain reaction, but in the Chernobyl reactor the graphite moderator was still in place. The speed of the chain reaction, and hence the rate at which it was producing heat, reached 100 times what the reactor was designed for.

Very high temperatures melt and vaporize things. This builds up pressure that can lead to damaging explosions. Intense heat also causes water to react with metals to form the explosive hydrogen gas we discussed in Chapter 6. No one knows exactly what took place inside the reactor, but at 1:24 A.M., there were two loud explosions, and glowing materials were seen flying out of the top of the reactor building. These explosions were not, of course, nuclear bomb detonations, but rather more like explosions of an overheated furnace or boiler. Nuclear bombs require much more highly enriched uranium than that used in power plants.

The most immediate problem at this point was to put out the fires, especially because there was another reactor in the same building that was in immediate danger. At about 1:30 A.M., firemen arrived from the nearby cities of Pripyat and Chernobyl, and by 3:54 A.M., the most threatening fires were out. By 5:00 A.M., all fires in the building were out and the other reactor was shut down. Two reactors in a contiguous building continued in operation for another 20 hours before receiving permission from Moscow to shut down. They were put back into operation a few months later.

The firemen displayed extraordinary heroism in putting out the fires. They received very high radiation doses, largely from radioactivity sticking to their bodies. In addition, they suffered thermal and chemical burns. Many of them later died. The most serious effects, due to beta rays irradiating their skin, would have been averted if they had worn the protective clothing routinely used in U.S. plants when dealing with radioactive contamination. In fact, it would have been very helpful even if attention had been paid to cleaning the exposed parts of their bodies to remove radioactive materials that were sticking to their skin.

The next problem was that the graphite inside the reactor was afire. Chemically, graphite is not much different from coal and is a fine fuel. Recall that radioactivity from a reactor is dangerous principally when it is dispersed into the environment as a fine dust. It would be difficult to devise a more efficient dispersal method than to engulf the radioactivity in a very hot fire sending a plume of smoke high up into the air. Radioactive dust was spewing out with this smoke. Something had to be done to put out the fire inside the reactor.

The firemen first attempted to pump water into the reactor, but that was unsuccessful. They then decided to drop materials from helicopters. From April 28 to May 2, 5,000 tons of boron compounds, dolomite, sand, clay, and lead were dropped onto the reactor. Boron was used because it strongly absorbs neutrons and thus stops a chain reaction. Lead melts easily, and they hoped it would flow over the top to keep air out and thus stop the burning. The dolomite, sand, and clay were to smother the fire. The helicopter pilots had to fly into the rising plume of radioactive dust, exposing many of them to heavy doses of radiation that later proved fatal. As a result of their heroism, the discharge of radioactive dust dropped sharply by May 6.

Many of the firemen and helicopter pilots, as well as some of the workers inside the plant, received radiation doses of more than a million millirems. In all, 31 men died, two of them killed immediately by the explosions, and the rest as a consequence of burns and radiation sickness (see Chapter 5). According to the group of Soviet physicians who toured the United States, none was saved by the bone marrow transplants carried out by a visiting American physician, accompanied by widespread publicity.

While these deaths among workers at the plant were horribly tragic, it is perhaps worth noting that an average of 50 deaths occur every day due to occupational accidents in the United States, and single accidents that kill more than 31 workers occur frequently in coal mines.

EFFECTS ON THE PUBLIC

There has been no direct evidence of injury to any member of the public as a result of the Chernobyl accident, but there were substantial doses of radiation. The city of Pripyat, with a population of 45,000, mostly families of plant workers, extends from near the plant to 2 miles away. However, exposure in that area averaged only 3,300 mrem, because radioactive materials projected high into the air did not descend rapidly enough to affect those close by. The largest exposures, averaging 50,000 mrem, were received by the 16,000 people who lived from 2 to 6 miles away. The 8,200 people living from 6 to 9 miles away also received substantial doses, averaging 35,000 mrem. The 65,000 people living from 9 to 18 miles away received only about 5,000 mrem. All of these 135,000 people were evacuated over the first few days to avoid further exposure from radioactive material deposited on the ground as well as from that still being released from the reactor.

For the most-exposed 16,000 who averaged 50,000 mrem, their risk of dying from cancer as a result is about 4%, raising their total risk of dying of cancer from the normal 20% to 24%. This is less than some of the variation in cancer risk from living in different U.S. states. A Soviet scientific team has announced plans to carefully keep track of these highly exposed people to determine how many cancers actually do appear.

For the first 2 days after the accident, the winds carried the radioactive dust over Finland and Sweden. On the third and fourth day, the wind shifted to bring it toward Poland, Czechoslovakia, Austria, and Northern Italy. It then shifted further southward to deposit the material over Rumania and Bulgaria.

People all over the world were exposed to external radiation from radioactive gases and dust suspended in the air and settled on the ground. They were also exposed internally by inhaling these materials or eating foods contaminated with them. The average radiation doses to the public[3] in millirems during the first year after the accident were 76 in Bulgaria, 67 in Austria, 40 to 60 in Greece, Rumania, and Finland, 30 to 40 in Yugoslavia, Czechoslovakia, and Italy, 20 to 30 in the USSR, Poland, Switzerland, Hungary, Norway, and East Germany, 10 to 20 in Sweden, West Germany, Turkey, and Ireland, and less than 10 elsewhere (0.15 in the United States and Canada). Note that in no country was the exposure higher than one-fourth of that due to natural radiation during that year.

Some of the material on the ground will continue to be radioactive for many years, exposing people externally and internally through the food supply. The estimated average total exposure in millirems after the first year[3] will be 120 in southeastern Europe, 95 in North and Central Europe, 81 in the USSR, 15 to 19 in Western Europe and Southwest Asia, 8 in North Africa, and less than 2 elsewhere (0.4 in North America). The sum of exposures to people all over the world will eventually, after about 50 years, reach 60 billion mrem,[3] enough to cause about 16,000 deaths. Note that this is still less than the number of deaths caused every year by air pollution from coal-burning power plants in the United States.

Since the mechanism for dispersing radioactivity over long distances was so efficient in the Chernobyl accident and is so inefficient in U.S. reactors, it is almost impossible to believe that an accident in a U.S. reactor can ever cause nearly as much radiation exposure at large distances from the plant.

LESSONS LEARNED

Any technology is developed and improved by learning from past mistakes, and the nuclear power establishment has always been very active in this regard. A tremendous amount was learned from the Three Mile Island accident. Within weeks, two large new organizations were established by the nuclear industry, the Nuclear Safety Analysis Center (NSAC) in Palo Alto, California, and the Institute of Nuclear Power Operations (INPO) in Atlanta, Georgia. Both have been very active since that time. NSAC carries out very intensive technical analyses, including some referred to in this book. INPO has established an impressive electronic communication network among power plants, reactor vendors, and centers of expertise all over the world. It also conducts inspections of power plants, offering suggestions and criticisms that are taken very seriously. As a result of INPO activities, there have been remarkable improvements in operational performance. The average number of operating hours per year for plants has increased by 12%. Unplanned shutdowns have been reduced by 70%, accident rates for workers have declined more than three-fold, the volume of low-level radioactive waste has declined by 72%, and radiation exposure to workers has been cut in half. INPO represents a very intense effort by the nuclear industry to police itself, as a result of the Three Mile Island accident.

Government agencies were also very active in trying to learn from the Three Mile Island accident. There was a Presidential Commission to investigate it and offer suggestions and criticisms of the industry. The NRC carried out very extensive studies and even produced a thick document entitled "Lessons Learned From The Three Mile Island Accident." The NRC also developed a set of modifications in nuclear plants to respond directly to some of the problems uncovered in the analysis of the accident; these modifications were quickly implemented at a cost of about $20 million per plant. In addition, the NRC undertook a broad study of a variety of nuclear safety issues. It is not an exaggeration to say that lessons learned from the Three Mile Island accident revolutionized the nuclear power industry.

After the Chernobyl accident, both government agencies and the nuclear industry were eager to investigate and learn from the experience. However, after long and careful study they finally concluded that we had very little to learn from it. The whole episode is now viewed as a vindication of the U.S. approach to nuclear power. (Essentially all nuclear power programs outside of the Soviet bloc use the U.S. approach.)

To understand this, let us review some of the problems that contributed to the accident, which would have been avoided by the U.S. approach:

1. A reactor which is unstable against a loss of water could not be licensed in the United States.
2. A reactor which is unstable against a temperature increase could not be licensed here.
3. A large power reactor without a containment could not be licensed here.
4. In contrast to the laxity at Chernobyl, regulations are strictly enforced here. Violations like operators cheating on examinations or falling asleep on the job, failing to report promptly on minor malfunctions, or failing to carry out a required inspection have brought large fines, plus lots of bad publicity to the utility. Flagrantly violat-

ing rules of reactor operation, and disabling important safety interlocks, are essentially unthinkable in U.S. plants.

5. U.S. commercial nuclear power plants would never be used for an experiment not directly relevant to their operations. Any abnormal use of the reactor would require very thorough advance analysis and approval from Washington based on government analyses. Major deviations from the plan, such as the decision at Chernobyl to run the experiment at a much lower power level, would never be considered. Presence of appropriate experts would be required. The preparations required to do such an experiment would be so massive that probably no one would even try. In the United States, experiments of this type are done at national laboratories.

6. In any severe accident in a U.S. reactor, the water would be lost, which means there would be no moderator. That would automatically stop the chain reaction. In a Chernobyl-type reactor, the moderator (graphite) is never lost, so there is no assurance that the chain reaction can be stopped in an accident. It could not be stopped in the April 1986 accident.

7. The principal way in which a reactor accident can do harm to public health is through dispersing the radioactive material as an airborne dust. A graphite fire is an efficient vehicle for doing this, provided by Chernobyl-type reactors. In essentially all accident scenarios for U.S. reactors, the bulk of the radioactive material ends up in water, and there is no reasonably probable mechanism for dispersing much of it as airborne dust.

Actually, it had been recognized for decades that Soviet reactors could not be licensed in the United States, even if they had containments. Items 4 and 5 in the above list further emphasize the fact that reactor safety has always received much lower priority in the Soviet Union than in the United States. An obvious question is why this is so.

There has been substantial contact between Soviet and American reactor safety experts, including many visits in both directions, personal friendships, and lots of informal discussions over cocktails. The above question has been asked and discussed many times, and the Soviet reply runs along the following lines.

The extreme concern about reactor safety in the United States has gone far beyond the bounds of rationality. It is difficult to argue with the Soviets on this point. In Chapter 8 we will show that our reactor safety programs have spent billions of dollars per expected life saved. This is irrational for two reasons. First there are many opportunities for saving lives with medical screening programs, highway safety measures, and the like, at a cost of about $100,000 per life saved, so the money spent to save one life from a nuclear reactor accident could save over 10,000 lives if spent in these other areas. Second, as a result of the cost increase for nuclear power plants, utilities are forced to build coal-fired power plants instead of nuclear plants, and the air pollution from a coal-fired plant is estimated to cause several thousand deaths over its operating lifetime. This irrational attitude toward nuclear reactor safety in the United States is, therefore, leading to thousands of unnecessary deaths every year, and wasting billions of dollars that could be used to save thousands of other lives. Why should the Soviet Union repeat our insanity?

It is difficult to argue with this logic of the Soviet reactor safety experts, but they carried things too far. The Chernobyl-type reactor is very much more dangerous than U.S. power reactors ever were. All the differences we have pointed out apply equally to even the earliest American power reactors. The United States has erred in one direction, but the Soviets erred in the other. They truly gambled, and they lost.

Chapter 8 / UNDERSTANDING RISK

One of the worst stumbling blocks in gaining widespread public acceptance of nuclear power is that the great majority of people do not understand and quantify the risks we face. Most of us think and act as though life is largely free of risk. We view taking risks as foolhardy, irrational, and assiduously to be avoided. Training children to avoid risk is an all-important duty of parenthood. Risks imposed on us by others are generally considered to be entirely unacceptable.

Unfortunately, life is not like that. Everything we do involves risk.[1] There are dangers in every type of travel, but there are dangers in staying home—25% of all fatal accidents occur there. There are dangers in eating—food is one of the most important causes of cancer and of several other diseases—but most people eat more than necessary. There are dangers in breathing—air pollution probably kills 100,000 Americans each year,[2] inhaling radon and its decay products is estimated to kill 14,000 a year[3], and many diseases like influenza, measles, and whooping cough are contracted by inhaling germs. These dangers can often be avoided by simply breathing through filters, but no one does that. There are dangers in working — 12,000 Americans are killed each year in job-related accidents, and probably 10 times that number die from job-related illness[4] — but most alternatives to working are even more dangerous. There are dangers in exercising and dangers in not getting enough exercise. Risk is an unavoidable part of our everyday lives.

That doesn't mean that we should not try to minimize our risks, but it is important to recognize that minimizing anything must be a quantitative procedure. We cannot minimize our risks by simply avoiding those we happen to think about. For example, if one thinks about the risk of driving to a destination, one might decide to walk, which in most cases would be much more dangerous. The problem with such an approach is that the risks we think about are those most publicized by the media, whose coverage is a very poor guide to actual dangers. The logical procedure for minimizing risks is to quantify all risks and then choose those that are smaller in preference to those that are larger. The main object here is to provide a framework for that process and to apply it to the risks in generating electric power.

There are many ways of expressing quantified risk, but here we will use just one, the loss of life expectancy (LLE); i.e., the average amount by which one's life is shortened by the risk under consideration. The LLE is the product of the probability for a risk to cause death and the consequences in terms of lost life expectancy if it does cause death. As an example, statistics indicate[5] that an average 40-year-old person will live another 37.3 years, so if that person takes a risk that has a 1% chance of being immediately fatal, it causes an LLE of 0.373 years (0.01×37.3).

It should be clear that this does not mean that he will die 0.373 years sooner as a result of taking this risk. But if 1,000 people his age took this risk, 10 might die immediately, having their lives shortened by 37.3 years, while the other 990 would not have their lives shortened at all. Hence, the average lost lifetime for the 1,000 people would be 0.373 years. This is the LLE from that risk.

Of course, most risks are with us to varying extents at all ages and the effects must be added up over a lifetime, which makes the calculations somewhat complex. We therefore developed a computer program for doing the calculations and used it to carry out a rather extensive study of a wide variety of risks. Some of the results of those studies are summarized in the next section.

A CATALOG OF RISKS[1,4,6]

One widely recognized risk is cigarette smoking. One pack per day has an LLE of 6.4 years for men and 2.3 years for women—in the former case, this corresponds to an LLE of 10 minutes for each cigarette smoked. For noninhalers, the lifetime risk from one pack per day is 4.5 years for men and 0.6 years for women, while for those who inhale deeply it is 8.6 years for men and 4.6 years for women. (The differences between male and female risks may involve how deeply they inhale, or some of their differences in lifestyle and physiology.) Giving up smoking reduces these risks: after 5 years the LLE is reduced by one-third, and after 10 years it is more than cut in half. Cigar and pipe smoking do little harm if there is no inhalation, but with deep inhalation the LLE is 1.4 years for pipes and 3.2 years for cigars.

Further understanding of the risks in smoking comes from examining the diseases from which smokers die more frequently. The following figures are death rate ratios for 1-2 packs per day, smokers to nonsmokers in the 35-84 age range. This ratio is 17 for lung cancer (i.e, heavy smokers are 17 times more likely to die from lung cancer than nonsmokers), 13 for cancer of the pharynx and esophagus, 6 for cancer of the mouth, 11 for bronchitis and emphysema, 4 for stomach ulcer, 3 for cirrhosis of the liver, 2 for influenza and pneumonia, 1.8 for cardiovascular disease, our nation's No. 1 killer, and between 1.5 and 2 for leukemia and cancer of the stomach, pancreas, prostate, and kidney.

Much attention has recently been given to the risk of being *near* a smoker. The best evidence on this is that the added risk of lung cancer for a nonsmoker due to being married to a smoker corresponds to an LLE of 50 days. There are probably additional risks due to other diseases, since only 13% of the deaths attributed to smoking are due to lung cancer.

Another major risk over which we have some personal control is being overweight—we lose about 1 month of life expectancy for each pound our weight is above average for our size and build. In the case of someone 30-pounds overweight, the LLE is thus 30 months, or 2½ years. To assess the effect of overeating, we note that our weight increases by 7 pounds for every 100-calorie increase in average daily food intake. That is, if an overweight man changes nothing about his eating and exercise habits except for eating one extra slice of bread and butter (100 calories) each day, he will gain 7 pounds (gradually over a period of about 1 year) and his life expectancy will be reduced by 7 months. This works out to a 15-minute LLE for each 100 extra calories eaten.

Any discussion of major risks must include the traditional leader—

disease. Heart disease leads in this category with an LLE of 5.8 years, followed in order by cancer, LLE 2.7 years; stroke, LLE 1.1 years for men and 1.7 years for women; pneumonia and influenza, LLE 4.5 months; and cirrhosis of the liver and diabetes, LLEs of a little over 3 months each, the former found more in men and the latter more in women by about 3 to 2 ratios.

One of the greatest risks in our society is remaining unmarried. Statistics show that a single white male has 6-years less life expectancy than a married white male; his LLE from being unmarried is thus 6.0 years. This figure is based on data obtained before AIDS appeared, so that is not involved here. One might suspect that part of the reason for these differences is that sickly people are less likely to marry, but this is evidently not the main reason, since mortality rates are even higher for widowed and divorced people at every age. The LLE for a white male (female in parentheses) from being unmarried at age 55 and not later marrying is 3.2 (1.8) years if he (she) is single, 3.9 (2.7) years if widowed, and 6.2 (2.5) years if divorced. For blacks these LLEs are 3.5 (4.1) years if single, 6.2 (6.0) years if widowed, and 6.0 (4.0) years if divorced. Note that males generally suffer more than females from remaining unmarried, and blacks suffer more than whites.

Unmarried people suffer excessively from a variety of diseases, and in general the widowed and divorced suffer from them more than those who have never married. For example, compared to married people of the same age, widowed males die more frequently from tuberculosis by 117%, from stomach and intestinal cancer by 26%, from lung cancer by 26%, from cancer of the genital organs by 23%, from leukemia by 8%, from diabetes by 41%, from stroke by 50%, from diseases of the heart and arteries by 46%, from cirrhosis of the liver by 142%, from motor vehicle accidents by 99%, from other accidents by 127%, from suicide by 139%, and from homicide by 69%. To mention a few extremes, compared to a married man of the same age, a divorced man is 6.1 times more likely to die of tuberculosis, 2.1 times more likely from lung cancer, 6.2 times more likely from cirrhosis of the liver, 3.8 times more likely from motor vehicle accidents, 4.2 times more likely from other accidents, 4.1 times more likely from suicide, and 7.2 times more likely from homicide.

Perhaps the least appreciated of all major risks is that of being poor, unskilled, and/or uneducated. The best data on the "unskilled" factor are based on occupational groupings. Professional, technical, administrative, and managerial people live 1.5 years longer than those engaged in clerical, sales, skilled, and semiskilled labor, and the latter group lives 2.4 years longer than unskilled laborers. Corporation executives live 3 years longer than even the longest-lived of the above groups, a full 7 years longer than unskilled laborers. A similar study in England showed even larger differences, finding comparable differences among wives of workers; the wife of a professional person lives about 4 years longer than the wife of an unskilled laborer. This indicates that the problems are not occupational exposures but rather are socioeconomic.

In seeking to understand the reasons for these differences, it is interesting to consider the causes of death. If we compare unskilled laborers with professional, technical, administrative, and managerial people in the United States, their risk of early death from tuberculosis is 4.2 times higher, from accidents 2.9 times higher, from influenza and pneumonia 2.8 times higher, from cirrhosis of the liver 1.8 times higher, and from suicide 1.7 times higher. It is also 30% higher from cancer and 13% higher from cerebrovascular disease, but it is 8% lower from arteriosclerosis and diabetes. The large factors in this list are from causes associated with unhealthy living conditions, limited access to medical treatment, or unenlightened attitudes toward health care, and are thus generally preventable. This would seem to be a fertile field for social action.

A similar pattern appears in correlations between life expectancy and educational attainment. College-educated people live 2.6 years longer than the average American, while those who dropped out of grade school live 1.7 years less than average, a 4.3-year differential. These differences are about the same for men and women, which indicates that occupational exposures are not the basic problem here. Dropping out of school at an early age ranks with taking up smoking as one of the most dangerous acts a young person can perform. Even volunteering for combat duty in wartime pales by comparison; the LLE from being sent to Vietnam during the war there was 2.0 years in the Marines, 1.1 years in the Army, 0.5 years in the Navy, and 0.28 years in the Air Force.

The data on poverty are truly impressive. In one study, the Chicago area was divided into sections based on socioeconomic class. The difference in life expectancy between the highest- and lowest-class sections was 9 years for white males, 7 years for white females, and nearly 10 years for nonwhites of both sexes. A Public Health Service survey in 19 U.S. cities found that mortality rates in poverty areas were typically 70% higher than in non-poverty areas, which corresponds roughly to a 10-year difference in life expectancy. Incidentally, in this survey the differences were substantially *less* than average for Chicago, where the detailed study described above was made.

There is also an abundance of data from foreign sources. In one study,

Montreal was divided into sections by socioeconomic class, and the difference in life expectancy between the highest- and lowest-class sections was 10.8 years for men and 7.3 years for women. A Canadian Ministry of Health Study in 21 cities divided the population into five equal parts (quintiles) by income, and found a difference in life expectancy between the highest and lowest quintiles to be 6 years for men and 3 years for women. (Note that the 20% with lowest income would hardly be classed as poverty cases.) A study in Finland divided the nation into four "social groups" and found a 7 year difference in life expectancy between the highest and lowest. A study in France found that professional and managerial men live 4 years longer than white collar workers, 9 years longer than skilled laborers, and 13 years longer than unskilled laborers.

Within technologically advanced nations there are effective programs to provide poor people with reasonable medical care and an adequate diet, but such programs are much fewer and less effective on an international basis. This leads to much larger variations of life expectancy with socioeconomic level, as evidenced by correlations between life expectancy and per capita gross national product for various nations.[7] In well-to-do countries like the United States, Western Europe, Australia, and Japan, life expectancy is about 75 years, whereas life expectancy in a sample of other countries is 72 years in Poland, Czechoslovakia, and Rumania; 67 years in Mexico and Central America; 64 years in Turkey, Brazil, and Thailand; 59 years in India, Iran, and Egypt; about 45 years in most central African countries; and 38 years in Afghanistan and Gambia. Lest these differences be ascribed to racial factors, it should be noted that Japanese have 10 years more life expectancy than other East Asians, and blacks in the United States have more than 20 years longer life expectancy than African blacks. The history of white versus black life expectancy in the United States is illuminating here: in 1900, there was a 18-year difference (50 years versus 32 years), whereas it is now reduced to 6 years, reflecting the improving socioeconomic status of blacks (but also showing that much progress remains to be made in that regard).

From all of this information, it is abundantly clear that wealth makes health, and poverty kills. Any action which might result in reducing our national wealth, or the wealth of segments of our population is fraught with danger. The LLEs we are dealing with here are about 10 years within the United States, and 30 years on an international basis.

Within the United States, life expectancy varies considerably with geography in ways not explainable by socioeconomic differences. For whites it is over a year longer than average in the rural north central states of North

Dakota, South Dakota, Minnesota, Iowa, Nebraska, Kansas, and Wisconsin (all of which have lower than the national average income), while it is a year shorter than average in the rural southeastern states of South Carolina, Georgia, Alabama, Mississippi, and Louisiana. It is speculated that this may be connected with trace elements in soil. Areas at higher elevation have a few years longer life expectancy than areas near sea level. The differences are especially large for cancer. Note that radiation exposure increases with altitude, which would tend to cause the reverse situation. The rocks at higher elevation generally contain more radioactive material, and there is less atmosphere above to shield out cosmic rays. This is another demonstration that radiation is not an important cause of cancer.

The most highly publicized risks are those of being killed in accidents — the suddenness and drama of accidental death are well suited to the functions of our news media — although the actual danger is well below that of the risks we have discussed previously. The LLE from all accidents combined is 435 days (1.2 years). Almost half of them involve motor vehicles, which give us an LLE of 180 days — 147 days while riding and 33 days as pedestrians. Using a small car rather than a large car roughly doubles one's risk, and this would be true even if everyone used small cars. The difference between using a small and a midsize car is an LLE of 60 days, and there is a roughly equal difference between a midsize and a large car. Reducing the national speed limit from 65 to 55 miles per hour in 1974 increased our life expectancy by 40 days, and the recent increase back to 65 miles per hour on interstate highways in some states has had a substantial reverse effect.

On an average, riding one mile in an automobile and crossing a street each have an LLE of 0.4 minutes, making them as dangerous as one puff on a cigarette (assuming 25 puffs to a cigarette), or, for an overweight person to eat three extra calories. Note that if walking involves crossing a street more often than once per mile, it is more dangerous to walk a mile than to drive a mile.

The total LLE over a lifetime from various other types of accidents is 40-days each for falls (mostly among the elderly) and drowning, 27 days for fire and burns, 17 days for poisoning, 13 days for suffocation, 11 days for accidents with guns, and 7½ days for asphyxiation. Men are more than twice as likely to die in accidents as women; in motor vehicle accidents the male/female ratio is 2½ to 1 for both riders and pedestrians, and in drowning the ratio is 5 to 1.

Accidental death rates vary greatly with geography; they are 4 times higher in Wyoming than in New York, to give the two extremes; the Northeast is generally the safest area, while the Rocky Mountain states are generally the most dangerous.

We spend most of our time at home and at work, so that is where most of our accidents occur that are not related to travel. The LLE for accidents in the home is 95 days, and for occupational accidents it is 74 days. The latter number varies considerably from industry to industry, from about 300 days in mining, quarrying, and construction to 30 days in trade (e.g., clerks in stores). Nearly half of all workers are in manufacturing and service industries, for which the LLE is 45 days. For radiation workers in the nuclear industry, radiation exposure gives them an average LLE of 24 days.

Actually, these statistics cover up many high-risk occupations because they average over whole industries including white collar workers and many others in relatively safe jobs. Canadian occupational accident statistics are kept in much finer detail and elucidate some of these effects:

- In the mining industry, the LLE for those who sink shafts is 660 days versus 65 days for those involved in shop work and service.
- In the utility industry, the LLE for those who work with power lines is 820 days versus 58 days for mechanics and fitters.
- In forestry, the LLE for those who fell trees is 1,050 days versus 54 days for sawmill workers.
- In construction, the LLE for demolition workers is 1,560 days (more than 4 years) versus 38 days for those involved with heating, plumbing, and electrical wiring.

Some showmanship activities are widely advertised as having very high accident potential, but judging from statistical experience, these dangers are exaggerated in the public mind. Professional aerialists—tight-rope walkers, trapeze artists, aerial acrobats, and high-pole balancers—get an LLE of 5 days per year of participation, or 100 days from a 20-year career. The risk is similar for automobile and motorcycle racers of various sorts. The risk of accidental death in these professions therefore is less than in ordinary mining and construction work. The most dangerous profession involving thousands of participants is deep-sea diving, with an LLE of 40 days per year of participation.

In addition to accidents, occupational exposure causes many diseases that affect a worker's life span that in most cases are much more important than accidents. Coal miners, on an average, die 3 years earlier than the average man in the same socioeconomic status, and statistics are similarly unfavorable for truckers, fishermen, ship workers, steel erectors, riggers, actors and musicians (perhaps due to irregular hours), policemen, and firemen. On the other hand, there are occupational groups in which men live a year or more longer than average for their socioeconomic standing, such as postal workers, government officials, university teachers, and gardeners. Clearly, one's choice of occupation can have a large effect on one's life expectancy, extending to several years.

But by far the most dangerous occupation is no occupation—being unemployed. For this we use a study by Ray Marshall, former Secretary of Labor and now a professor at the University of Texas. Unemployment affects not only the worker himself, but his family and friends, and even those who remain employed because of stress caused by fear of losing their jobs. But if all of these effects were concentrated on the worker himself, the LLE from one year of unemployment would be about 500 days. This is about equal to the risk of smoking 10 packs of cigarettes per day while unemployed.

The unemployment rate in the United States frequently rises or falls by 1% or more. The estimated effects of a 1% increase for one year are 37,000 deaths, including 20,000 due to cardiovascular failure, 500 due to alcohol-related cirrhosis of the liver, 900 suicides, and 650 homicides. In addition to the deaths, there are 4,200 admissions to mental hospitals and 3,300 admissions to state prisons. Clearly, any action, or inaction, that can lead to increased unemployment is very dangerous. Importing oil rather than utilizing domestic energy production is such an action, and having inadequate supplies of electricity or allowing electricity costs to rise unnecessarily are such inactions.

Medical care is an obvious factor affecting life expectancy. If full use were made of available medical technology, it is estimated that 75,000 cancer deaths and 125,000 deaths from cardiovascular diseases could be prevented each year. Failure to achieve this performance by our medical care system is costing the average American an LLE of about 1.4 years.

It is estimated that if all currently available technology were used to prolong life, including good dietary practice, proper exercise and rest, and best available medical care, life expectancy would be increased by 9.5 years. Thus, suboptimal lifestyles give us an LLE of 9.5 years. Over 20% of this is due to cigarette smoking.

Averaged over the U.S. population, AIDS gives an LLE of 70 days. That disease is preventable by careful sexual practices.

Homicide and suicide are significant risks in our society, with LLEs of about 135 days each for men, and 43 and 62 days, respectively, for women. Homicide is more common among the young, while suicide becomes several times more likely among the elderly.

Judging by the media coverage they attract, one might think that large catastrophes pose an important threat to us, but this is hardly the case. Hurricanes and tornadoes combined give the average American an LLE of one day, as do airline crashes. Major fires and explosions (those with eight or more fatalities) give us an LLE of 0.7 days, and our LLE from massive chemical releases is only 0.1 day.

The media have publicized the dangers of various individual substances from time to time. Broiling meat converts some of it into carcinogenic compounds, and burning charcoal which is sometimes used in broiling adds others; eating a half-pound of broiled steak per week throughout life gives an LLE of 0.15 days. Peanut butter contains aflatoxin, which causes liver cancer; one tablespoon per day gives us an LLE of 1.1 days. Milk also contains aflatoxin, enough to cause an LLE of 1.0 day from drinking one pint per day. Chlorination of drinking water leads to formation of chloroform, which is a carcinogen; in Miami or New Orleans where chlorination levels are particularly high, this gives an LLE of 0.6 days. Of course, chlorination of water and drinking milk have benefits that far outweigh these risks. There is evidence that coffee can cause cancer of the pancreas, giving an LLE of 26 days from drinking 2½ cups per day, the U.S. average. Birth control pills can cause phlebitis, giving an LLE of 5 days to users. Diet drinks with saccharin, which causes bladder cancer, give an LLE of 2 days from one 12-ounce serving per day, but this is a hundred times less harmful than the weight gain from nondiet drinks if one is overweight.

Other things some of us use are much more harmful. Alcohol abuse is estimated to cause 100,000 deaths per year due to cirrhosis of the liver, psychosis, accidents, suicides, and homicides, giving the average American an LLE of 230 days. Abuse of other addictive substances is estimated to cause 35,000 deaths per year, corresponding to an LLE of 100 days averaged over the population. Of course, the LLE is much larger than these averages for those who indulge and much less for those who do not. However, it is still substantial for the latter group, as they are often victims of homicides and of automobile collisions with drunk drivers.

Even very tiny risks often receive extensive publicity. Perhaps the best example was the impending fall of our orbiting Sky-Lab satellite, which gave us an LLE of 0.002 seconds. Heavy publicity surrounded leaks from radioactive waste burial grounds, although these have not given any single member of the public an LLE as large as 10 seconds. It is shown in the Chapter 8 Appendix that the Three Mile Island nuclear power plant accident gave the average Harrisburg-area resident an LLE of 2 minutes (0.0015 days). Our risk of being struck by lightning gives us an LLE of 20 hours.

Football seems like a dangerous sport, but the risk of being killed per year of participation is only 1 in 81,000 in high school and 1 in 33,000 in college, corresponding to LLEs of 0.3 and 0.6 days, respectively. Many other sports are much more dangerous. The LLE per year of participation is 8 days for professional boxing, 25 days for hang gliding, 110 days for dedicated mountain climbers (10 days for all climbers), 0.9 days for mountain hikers, 25 days for parachuting, 9 days for sailplaning, 7 days for amateur scuba diving, 2 days for snowmobiling, and 0.5 days for racing on skis.

There are several very large risks that are so mundane that we often ignore them. Females live nearly 8 years longer than males, and whites live 5.5 years longer than blacks. One might therefore say that the LLE from being a male rather than a female is 8 years, larger than for almost any other risk we have considered, and the LLE due to being black is 5.5 years, although much of this may be due to socioeconomic factors.

For convenient reference, some of the LLEs we have discussed are summarized in Table 1, and shown graphically in Fig.1.

RISKS OF NUCLEAR ENERGY IN PERSPECTIVE

With the benefit of this perspective, we now turn to the risks of nuclear energy, and evaluate them as if a large fraction of the electricity now used in the United States were generated from nuclear power. The calculations are explained in the Chapter 8 Appendix, but here we will only quote the results.

According to the Reactor Safety Study by the U.S. Nuclear Regulatory Commission (NRC) discussed in Chapter 6, the risk of reactor accidents would reduce our life expectancy by 0.012 days, or 18 minutes, whereas the antinuclear power organization Union of Concerned Scientists (UCS) estimate is 1.5 days. Since our LLE from being killed in accidents is now 400 days, this risk would be increased by 0.003% according to NRC, or by 0.3% according to UCS. This makes nuclear accidents tens of thousands of times less dangerous than moving from the Northeast to the West (where accident rates are much higher), an action taken in the last few decades by millions of Americans with no consideration given to the added risk. Yet nuclear accidents are what a great many people are worrying about.

The only other comparably large health hazard due to radiation from the nuclear industry is from radioactivity releases into the environment during routine operation (see Chapter 12). Typical estimates are that, with a full nuclear power program, this might eventually result in average annual exposures of 0.2 mrem (it is now less than one-tenth that large), which would reduce our life expectancy by another 37 minutes (see Chapter 8 Appendix).

TABLE 1
LOSS OF LIFE EXPECTANCY (LLE) DUE TO VARIOUS RISKS[a]

Activity or risk	LLE (days)
Living in poverty	3500
Being male (vs. female)	2800
Cigarettes (male)	2300
*Heart disease	2100
Being unmarried	2000
Being black (vs. white)	2000
Socioeconomic status, low	1500
Working as a coal miner	1100
*Cancer	980
30-lb overweight	900
Grade school dropout	800
*Suboptimal medical care	550
*Stroke	520
15-lb overweight	450
*All accidents	400
Vietnam army service	400
Living in Southeast (SC,MS,GA,LA,AL)	350
Mining construction (accidents only)	320
*Alcohol	230
Motor vehicle accidents	180
*Pneumonia, influenza	130
*Drug abuse	100
*Suicide	95
*Homicide	90
*Air pollution	80
Occupational accidents	74
*AIDS	70
Small cars (vs. midsize)	60
Married to smoker	50
*Drowning	40
*Speed limit: 65 vs. 55 miles per hour	40
*Falls	39
*Poison + suffocation + asphyx	37
*Radon in homes	35
*Fire, burns	27
Coffee: 2½ cups/day	26
Radiation worker, age 18–65	25
*Firearms	11
Birth control pills	5
*All electricity, nuclear (UCS)	1.5
Peanut butter (1 Tbsp./day)	1.1
*Hurricanes, tornadoes	1
*Airline crashes	1
*Dam failures	1
Living near nuclear plant	0.4
*All electricity, nuclear (NRC)	0.04

[a]Asterisks indicate averages over total U.S. population; others refer to those exposed.

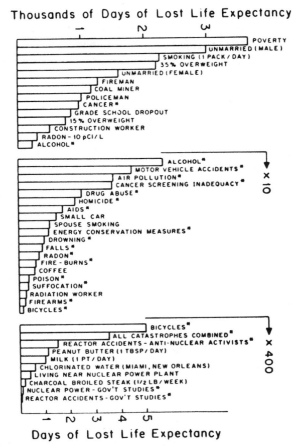

Fig. 1. Bar diagram for comparing risks. Height of bar is LLE. Asterisk designates average risk spread over total U.S. population; others refer to risks of those exposed or participating.

This brings the total from nuclear power to about 1 hour (with this 37 minutes added, the UCS estimate is still about 1.5 days).

If we compare these risks with some of those listed in Table 1, we see that having a full nuclear power program in this country would present the same added health risk (UCS estimates in brackets) as a regular smoker indulging in one extra cigarette every 15 years [every 3 months], or as an overweight person increasing his or her weight by 0.012 [0.8] ounces, or as in raising the U.S. highway speed limit from 55 miles per hour to 55.006 [55.4] miles per hour, and it is 2,000 [30] times less of a danger than switching from midsize to small cars. Note that these figures are not controversial, because I have given not only the estimates of Establishment scientists but also those of the leading nuclear power opposition group in this country, UCS.

I have been presenting these risk comparisons at every opportunity for several years, but I get the impression that they are interpreted as the opinion of a nuclear advocate. Media reports have said "Dr. Cohen claims..." But there is no personal opinion involved here. Deriving these comparisons is simple and straightforward mathematics which no one can question. I have published them in scientific journals, and no scientist has objected to them. I have quoted them in debates with three different UCS leaders and they have never denied them. If anyone has any reason to believe that these comparisons are not valid, they have been awfully quiet about it.

It seems to me that these comparisons are the all-important bottom line in the nuclear debates. Nuclear power was rejected because it was viewed as being too risky, but the best way for a person to understand a risk is to compare it with other risks with which that person is familiar. These comparisons are therefore the best way for members of the public to understand the risks of nuclear power. All of the endless technical facts thrown at them are unimportant and unnecessary if they only understand these few simple risk comparisons. That is all they really need to know about nuclear power. But somehow they are never told these facts. The media never present them, and even nuclear advocates hardly ever quote them. Instead, the public is fed a mass of technical and scientific detail that it doesn't understand, which therefore serves to frighten.

When I started my investigations into the safety of nuclear energy in 1971, I had no preconceived notions and no "axes to grind." I was just trying to understand in my own way what the fuss was all about. Rather early in these efforts, I started to develop these risk comparisons. They convinced me that nuclear power is acceptably safe with lots of room to spare. If I am a nuclear advocate, it is because developing these comparisons has made me so.

To be certain that this all-important bottom line is not missed, let me review it. According to the best estimates of Establishment scientists, having a large nuclear power program in the United States would give the same risk to the average American as a regular smoker indulging in one extra cigarette every 15 years, as an overweight person increasing his or her weight by 0.012 ounces, or as raising the U.S. highway speed limit from 55 to 55.006 miles per hour, and it is 2,000 times less risky than switching from midsize to small cars. If you do not trust establishment scientists and prefer to accept the estimates of the Union of Concerned Scientists, the leading nuclear power opposition group in the United States and scientific advisor to Ralph Nader, then having all U.S. electricity nuclear would give the same risk as a regular smoker smoking one extra cigarette every 3 months, or of an overweight person increasing his weight by 0.8 of an ounce, or of raising the U.S. highway speed limit from 55 to 55.4 miles per hour, and it would still be 30 times less risky than switching from midsize to small cars. The method for calculating these numbers is explained in Chapter 8 Appendix.

Some people think of radiation risks as somehow being different from other risks. Maybe they don't believe that scientists really understand health effects of radiation, or perhaps they view death from radiation as worse than death from other causes. For these people it is more meaningful to compare risks of nuclear power with another radiation risk, radon in homes. The average American gets an LLE of 35 days from radon, 800 times as much as from nuclear power (23 times as much according to UCS). Exposure from radon can be greatly reduced by opening a window, or by being outdoors. Thus, the *radiation* risk of nuclear power is equal to that of staying home an extra 8 hours per year, or of keeping windows closed an average of 2 minutes extra per day. There are also simple things one can do to drastically reduce radon levels, such as opening two windows in the basement and putting a fan in one of them. Even in cold climates this can be done with no problems for 6 months of the year, which would be 10 times more effective than eliminating nuclear power, even if one accepts the risk estimates of UCS (400 times more effective according to most estimates).

All of this applies to homes with average radon levels, but several million Americans live in homes with over 10 times this average. For them, reducing radon exposure would be more than 10 times as effective. For example, their radiation exposure from nuclear power is equal to that of spending an extra 45 minutes per year indoors or of keeping their windows closed an average of an extra 12 seconds per day.

The strange part of all this is that, for all of its concern about the radiation from nuclear power, the public has shown little concern about radon. Only 2% of Americans have even bothered to measure the radon level in their homes, despite repeated statements from the Surgeon-General's office and the Environmental Protection Agency that it should be measured in every home.

The purpose of the discussion presented above is to make the risks of nuclear power understandable. Risks are best understood when compared to other risks with which we are familiar. But we have not discussed the question of whether they are acceptable. Acceptability includes other factors than the magnitude of risks. People are more willing to accept voluntary risks like skiing, auto racing, and mountain climbing than involuntary risks like pollution. Opponents are quick to point out that nuclear power presents an involuntary risk to the public. On the other hand, most of the other risks we have discussed are involuntary or at least have an involuntary component. Poor people certainly are not poor by choice. Living in Southeastern states is no more voluntary than living near a nuclear plant. In many if not most cases, a person's occupation is determined more by circumstances than by voluntary choice; a boy who grows up in the coal fields often has not been exposed to other occupational options and has little opportunity to explore them, so his becoming a coal miner is far from voluntary.

Riding in automobiles is hardly voluntary for most people, as they have no other way to get to work, to purchase food, and to participate in other normal activities of life; even if one avoids riding in automobiles, one is still subject to accidents to pedestrians, which account for 20% of deaths from motor vehicle accidents. A large fraction of other accidents are largely due to involuntary activities. Most drownings are of children, but a parent cannot prevent her child from going swimming without risking psychological damage to the child. An appreciable number of drownings result from taking baths, which is hardly a voluntary activity. Deaths from fire, burns, falls, poison, suffocation, and asphyxiation are also not usually due to voluntary risk taking.

Some people are more willing to accept natural risks than manmade risks, but nearly all of the risks we have considered are manmade. Living with artificial risks is part of the price we pay for the benefits of civilization.

Some say that risks which occur frequently but kill only one or a few people at a time are far less important than occasional large catastrophes which kill the same total number of people. That is the prima facie attitude of the media since their coverage is certainly much greater for the latter than for the former. Based on this viewpoint, some people attribute greater importance to the very rare reactor meltdown accident in which there might be numerous deaths, than to air pollution which kills far larger numbers of people one at a time.

Actually this argument is highly distorted. The cancers from even the worst meltdown accident considered in the Reactor Safety Study (RSS) would not be any more noticeable than deaths now occurring from air pollution; it would increase the cancer risk of those exposed by only about half of one percent, whereas normal cancer risks are 20%, varying by several percent with geography, race, sex, and socioeconomic status. Noticeable fatalities are expected in only 2% of all meltdowns, and even the worst meltdown treated in the RSS is expected to cause only a few thousand such deaths. There has already been a comparable disaster from coal burning, an air pollution episode in London in 1952 in which there were 3,500 more deaths than normally expected within a few days.

Since reactor meltdowns are *potential* accidents, none having occurred, they should be compared with other potential accidents. There are dam failures which could kill 200,000 people within a few hours; they are estimated to be far more probable than a bad nuclear meltdown accident. There are many potential causes of large loss of life anywhere large numbers of people congregate. A collapse of the upper tier of a sports stadium, a fire in a crowded theater, and a poison gas getting into a ventilation system of a large building (some buildings house 50,000 people) are a few examples. Any of these are far more likely than the fraction of one percent of all nuclear meltdowns that would cause large loss of life. The idea that a potential reactor meltdown accident is uniquely or even unusually catastrophic is grossly erroneous.

But I have a deeper objection to the idea that a catastrophic accident is more important than a large number of people dying unnoticeably. To illustrate, suppose there are two technologies, A and B, for producing a desired product, that Technology A causes one large accident per year in which 100 people are killed, and that Technology B kills 1,000 people each year one at a time and unnoticeably. If Technology B is chosen to avoid the catastrophic accidents from Technology A, 900 extra people die unnecessarily each year. How would you like to explain to these 900 people and their loved ones that they must die because the public is more interested in large catastrophes? Moreover, any one of us might be one of these 900 that die unnecessarily each year. I therefore maintain that in choosing between technologies on the basis of health impacts, the total number of deaths or the LLE should be the overriding consideration.

If we are forced to accept the overblown importance of catastrophic accidents, at the very least it should be clearly recognized that sensationalized media coverage with no attempt to put risks into perspective is responsible for large numbers of needless deaths.

What risks are acceptable is not a scientific question, and I as a scientist therefore cannot claim expertise on it. I have merely represented the risks as they are, I hope in understandable terms. If any citizen feels that the benefits of electricity produced by nuclear power plants are not worth the risk to the average citizen of a regular smoker smoking one extra cigarette every several years, or of an overweight person adding a fraction of an ounce to his weight, or of raising the national speed limit from 55 to something like 55.1 miles per hour, he is entitled to that opinion. However, he is obligated to suggest a substitute for the nuclear electricity.

Nearly all experts agree that the most viable substitute is coal-burning power plants. In Chapter 3 we noted that air pollution is estimated to be causing about 100,000 deaths per year. These are generally among the sick and elderly who would probably have died within a few years without the air pollution. On the other hand, exposure to air pollution throughout their lives has contributed to their sickly condition. We therefore judge that lifelong exposure to air pollution has reduced the life span of an average victim by about 7 years. On this basis, it gives the average American an LLE of about 80 days.

In order to determine what fraction of the LLE is due to coal-burning power plants, we would have to know what components of air pollution are responsible for these deaths. The two principal sources of air pollution are coal burning and automobiles. The latter is important principally in areas with lots of sunlight, which is needed to convert the nitrogen oxides and hydrocarbons into damaging materials like ozone and PAN. This applies to cities like Los Angeles or Phoenix, and these places generally do little coal burning. All of the killer episodes mentioned in Chapter 3 were in areas where air pollution was dominated by coal burning, and the same is true of the cross sectional and time series studies on which the estimate of deaths is based. In fact, there is little evidence that Los Angeles smog, for all the discomfort and unpleasantness it causes, is responsible for any deaths. There can therefore be little doubt but that coal burning is responsible for most of the 100,000 deaths per year. Since somewhat over half of the coal burning is in power plants, we estimate that air pollution from coal-burning power plants is responsible for about 30,000 deaths per year, which means that it gives the average American an LLE of 30 days. It also means that a large coal-burning plant causes something like 70 deaths per year, or 3,000 deaths over its operating lifetime.

This estimate is highly uncertain. It could easily be twice as high or one-third as high. Opponents of nuclear energy have constantly used the uncertainty in radiation risks from nuclear power to imply that things may be much worse than usually estimated; they get a lot of mileage out of this ploy, but the uncertainties in health effects of air pollution from coal burning are very much larger.

In any case, there can be little question but that the LLE from air pollution due to coal burning is considerably larger than that due to all the radiation and accidents from the nuclear industry even if the claims of the nuclear power opponents are accepted—30 days versus 1.5 days. According to the much more widely accepted estimates, the ratio of coal burning to nuclear LLE is 30 days versus one hour, a ratio of more than 700.

During the 1970s, there were numerous studies of the comparison of health impacts between nuclear and coal-burning electricity generation. At that time, effects of air pollution were estimated to be considerably less serious and much more easily curable by simply reducing sulfur dioxide releases from coal burning. Moreover, the effectiveness of the emergency core cooling system in preventing a reactor meltdown was still in doubt, and the results of the Reactor Safety Study were still new and unverified by independent investigations. Nevertheless, every one of these studies concluded that nuclear power was less harmful to human health than coal burning.[8] I know of no study that reached the opposite conclusion, that nuclear power is more harmful. In 1981, I published an offer of a $50 reward for information leading to my discovery of such a study, and that offer has been repeated several times in publications and over a hundred times in speaking engagements. All anyone would have to do to claim the reward is let me know how to get a copy of the report of such a study. Nevertheless, there have been no takers. I can only conclude that there are no studies which conclude that nuclear power is more harmful than coal burning.

The fact that nuclear power is less harmful to health than coal burning was conceded in a report by Union of Concerned Scientists, the leading nuclear power opposition organization, and by its president, Henry Kendall, in a scientific seminar at Carnegie-Mellon University. When I asked Ralph Nader about this, his reply was "maybe we shouldn't burn coal either." He didn't offer an alternative. Perhaps he thinks we don't need electricity.

Other than coal burning, the principal alternatives to nuclear power are oil, gas, solar energy, and conservation, i.e., reducing our energy use. We consider only their health impacts here, although these are by no means their most important drawbacks, and some of their other problems are discussed elsewhere in this book. Burning oil causes some air pollution, although not as much as coal burning; it is also responsible for fires, giving a total LLE of 4 days. Natural gas causes a little air pollution and some fires, but also kills by explosions and asphyxiation, giving an LLE of 2½ days. The principal health impact of solar energy is in the coal that must be burned to produce the vast quantities of steel, glass, and concrete required to emplace the solar collectors; this is about 3% of the coal that would be burned to produce the same energy by direct coal burning,[9] so the health effects are 3% of those of the latter, or an LLE of 1.0 day, if we obtained all of our electricity from the sun. This makes solar electricity far more dangerous to health than nuclear energy according to estimates by most scientists. The quantities of material used in other technologies are many times less than those required for solar technologies.

All electrical energy technologies bring with them the risk of electrocution, which has an LLE of 5 days for the average American. Note that this is far higher than the effects of generating nuclear electricity even if we accept the estimates of the nuclear power opponents. If solar electricity is generated and power conditioned in homes, it would probably multiply this effect manyfold.

The final alternative to nuclear power is conservation, doing without so much energy. Improvements in efficiency are, of course, always welcome, and there has been heartening progress on this in recent years. Waste is bad by definition. But while many people think that doing without energy is the safest strategy, it is probably by far the most dangerous. One energy conservation strategy is to use smaller cars, but we have shown that this has an LLE of 60 days, many times that of any other energy technology. The danger would be somewhat reduced if everyone used small cars, but most fatal accidents are from collisions with fixed objects, like trees or walls, and collisions with large trucks and buses are also important; the risk they pose is greatly increased by using small cars.

If fuel conservation doubles the amount of bicycling, the resulting LLE is 10 days for the average American. Bicycles are far more dangerous than small cars, and motorcycles are in between, both in danger and in fuel conservation. Another energy conservation strategy is to seal buildings more tightly to reduce the escape of heat, but this traps unhealthy materials like radon inside. Tightening buildings to reduce air leakage in accordance with government recommendations would give the average American an LLE of about 20 days due to increased radon exposure, making conservation by far the most dangerous energy strategy from the standpoint of radiation exposure!

Still another conservation strategy is to reduce lighting. Falls now give us an LLE of 39 days; thus if reduced lighting causes 5% more falling, it has an LLE of 2 days. If reduced lighting increases the number of murders by 5%, this would give an additional LLE of 4.5 days. Most motor vehicle accidents occur at night, although most driving is done during the day; if reduced road lighting increases accidents by even 2%, this gives an additional 4 days of LLE. Note that each of these LLEs is larger than that due to nuclear power even if the estimates of its opponents are accepted.

An important potential danger in overzealous energy conservation is that it may reduce our wealth by suppressing economic growth. Just to keep up with our increasing population without increasing unemployment, we must provide over a million new jobs per year for the foreseeable future. We have been succeeding in keeping up for the past several years, but that requires increasing supplies of electricity. The tragic health impacts of unemployment were outlined earlier in this chapter.

We have also noted the health effects of a nation's wealth. Typically, life expectancy is 75 years in economically advanced nations versus 45 years in poor nations, a difference of 30 years.

From this it is evident that reduced national wealth can have disastrous impacts on health, amounting to hundreds of days of LLE. The greatest potential risk in overzealous energy conservation is that it may lead us down that thorny path toward becoming a poorer nation.

Since we have estimated the LLE from coal burning as only 30 days, that technology is much less risky than doing without the electricity we need. Coal burning is an acceptable method for generating electricity, but is far inferior to nuclear energy for that purpose.

Some people seem to believe that reducing our energy usage is unavoidable because we are running out of fuel, but that is most definitely not the case. We will show in Chapter 13 that nuclear fission can easily provide all the energy the world will ever need without any increase in fuel costs. No one favors wasting energy; waste is bad by definition. But there is no long-term reason to deny ourselves any convenience, comfort, or pleasure that energy can bring us, as long as we are willing to pay a fair market price for it.

SPENDING MONEY TO REDUCE RISKS[10,11]

Another aspect of understanding risk is to consider what we are doing—or deciding not to do—to reduce our risks. Surely it is unreasonable to spend a lot of money to reduce one risk if we can much more cheaply reduce a greater risk but are not doing so.

It may seem immoral and inhumane even to consider lifesaving in terms of money, but the fact is that a great many of our risks can be reduced by spending money. In the early 1970s, air bags were offered as optional safety equipment on several types of automobiles, but this was discontinued because not enough people were willing to buy them. They were proven to be effective and safe—an estimated 15,000 lives per year would be saved if they were installed in all cars. There is no discomfort or inconvenience connected with them. They have only one drawback—they cost money. Apparently

Americans did not feel that it was worth the money to reduce their risk of being killed or injured in an automobile accident.

There is a long list of other automobile safety features. We can buy premium tires, improved lights, and antilock braking systems, to name a few. We can spend money on frequent medical examinations, we can use only the best and most experienced doctors, we can buy elaborate fire control equipment for our homes, we can fly and rent a car at our destination rather than drive on long trips, we can move to safer neighborhoods — the list is endless. Each of these also costs money. In this section we consider how much it costs to save a life by spending money in various ways. In some cases, where personal effort and time are also required, a reasonable monetary compensation for these will be added to the cost.

As an example, getting a Pap smear to test for cervical cancer requires making an appointment and going to the doctor's office; most women would be willing to do equivalent chores for a payment of $10. A Pap test costs about $20, so we add the $10 for time and effort and take the total cost to be $30. Each annual Pap test has 1 chance in 3,000 of saving a woman's life; thus for every 3,000 tests, costing (3,000 × $30 =) $90,000, a life is saved. The average cost per life saved is then $90,000. About 50% of U.S. women of susceptible age now have regular Pap tests. If you are among the other 50%, you are effectively deciding that saving your life is not worth $90,000.

This example is taken from a study completed a few years ago,[9] in which all costs are given in 1975 dollars. Some other calculational details will be given in the Chapter 8 Appendix, but here we will quote some of the results of that study. Since the consumer price index has roughly doubled since 1975, we will double all costs given in the original paper.

If there were smoke alarms in every home, it is estimated that 2,000 fewer people would die each year in fires. Even with a generous allowance for costs of installation and maintenance, this works out to a life saved for every $120,000 spent, but less than half of American homes have smoke alarms.

On the other hand, a great many Americans purchase premium tires to avert the danger of blowouts. If everyone did, this would cost an aggregate of about $10 billion per year and might avert nearly all of the 1,800 fatalities per year that result from blowouts, a cost of nearly $6 million per life saved. Many Americans buy larger cars than they need in order to achieve greater safety, which costs something like $12 million per life saved.

There is clearly no logical pattern here. It is not that some people feel that their lives are worth $12 million while others do not consider theirs to be worth even $90,000 — there are undoubtedly many women who buy larger cars for safety reasons but skip their regular Pap test. And there are millions of Americans who purchased premium tires with their new cars but did not order air bags, even though the air bags are 10 times more cost effective. The problem is that the American consumer does not calculate cost effectiveness. His or her actions are governed by advertising campaigns, salesmanship, peer group pressures, and a host of other psychological and sociological factors.

But what about the government? We pay a large share of our income to national, state, and local governments to protect us. The government can hire scientists or solicit testimony from experts to determine risks and benefits, or even to develop new methods for protecting us; they have the financial resources and legal power to execute a wide variety of health and safety measures. How consistently has the government functioned in this regard?

First let us consider cancer-screening programs. The government could implement measures to assure that much higher percentages of women get annual Pap smears; this has been done in a few cities like Louisville, Toledo, Ostfold (Norway), Aberdeen (Scotland), and Manchester (England). Ninety percent participation was achieved by such measures as sending personal letters of reminder or visits by public health nurses. Such measures would involve added costs, but tests would be cheaper when done in a large-scale program—a Mayo Clinic program did them for $3.50 in the 1960s and a British program did them for $2 apiece in 1970. Thousands of lives could be saved each year at a cost below $100,000 each.

There are several other cancer-screening programs that could be implemented. Fecal blood tests can detect cancer of the colon or rectum for as little as $20,000 per life saved. Many more of these cancers could be detected in men aged 50-65, the most susceptible age, by annual proctoscopic examinations, saving a life for every $60,000 spent, but only one in eight men of this age now get such examinations. Lung cancer can be detected by sputum cytology and by X-ray examination; the Mayo Clinic has been running such a program for heavy cigarette smokers that saves a life for every $130,000 spent, and two programs in London reported success at less than half of that cost. Nevertheless, only a small fraction of American adults are screened each year, and there has been little enthusiasm for large government-sponsored programs. Estimated costs per life saved are similar for breast cancer—the leading cause of cancer death in women—which is readily curable with early detection, but only half of all women are serviced, and again there are no large government programs.

Testing for high blood pressure has almost become a fad in this country, but the problem goes beyond detection. Treatment is quite effective, but since the condition is not immediately life threatening, many people ignore it. A well-organized treatment program would save a life for every $150,000 invested, with half of that cost compensating patients for their inconvenience, but such programs have not been developed.

An especially effective approach to saving lives with medical care is with mobile intensive care units (MICUs), well-equipped ambulances carrying trained paramedics ready to respond rapidly to a call for help. About one-third of all deaths in the United States are from heart attacks (one-third of these are in people less than 65 years old) and two-thirds of these deaths occur before the patient reaches the hospital. The MICU was originally conceived as a method for combating this problem by providing rapid, on-the-scene coronary monitoring and defibrillation services, but it has now been expanded to provide treatment for burns, trauma, and other emergency conditions. Experience in large cities has shown that MICUs save lives for an average cost of about $24,000; consequently, every large city has them. However, for smaller towns the cost goes up. When it reaches $60,000 per life saved—the cost for a town with a population of 40,000—MICU is often considered too expensive. In effect, it is decided that saving a life is not worth more than $60,000.

To summarize our medical examples, there are several available programs that could save large numbers of lives for costs below $100,000 each, and many more for costs up to $200,000 per life saved. These, of course, are American lives, with some chance that they may be our own.

There are numerous opportunities for highly cost-effective lifesaving in underdeveloped countries. The World Health Organization (WHO) estimates that over 5 million childhood deaths could be averted each year by immunization programs at a cost ranging from $50 per life saved from measles in Gambia and Cameroon to $210 per life saved by a combination of immunizations in Indonesia. These costs are for complete programs that provide qualified doctors and nurses, medical supplies, transportation, communication, and the like. WHO also estimates that about 3 million childhood deaths each year could be averted by oral rehydration therapy (ORT) for diarrhea. This consists of feeding a definite mixture of salt, sugar, baking soda, and "sodium-free" salt with water on a definite schedule. The cost per life saved by complete programs range from $150 in Honduras to $500 in Egypt.

Other low-cost approaches to lifesaving in the Third World include malaria control ($550/life saved), improved health care ($1930), improved water sanitation ($4030), and nutrition supplements to basic diets ($5300).

But health care is not our government's only means of spending money to save lives. Over 35,000 Americans die in automobiles each year as a result of collisions, and over a million are seriously injured, even though there are many ways in which this toll could be reduced by investing money in highway or automobile safety devices.

To some extent this has been done. A number of new safety devices in automobiles, like collapsible steering columns and soft dashboards, were mandated by law between 1965 and 1974; a study by the U.S. General Accounting Office indicated that they have saved a life for every $280,000 spent. However, this is apparently about as high as we are willing to go, and the program has ground to a halt. In 1970-1973, 16 new safety measures in automobiles were mandated, but there have been hardly any since that time. As noted previously, an air bag requirement that would cost $600,000 per life saved has not been implemented.

There are lots of highway construction measures that could save lives. For example, about 6,000 Americans die each year in collisions with guardrails, and there are guardrail construction features that could save most of those lives. But let us now get down to costs per life saved.

According to a recent report by the U.S. Department of Transportation, current programs save 79 lives per year with improved traffic signs, at a cost of $31,000 per life saved; 13 lives per year with improved lighting, for $80,000 per life saved; 119 lives per year with upgraded guardrails, for $101,000 per life saved; 28 lives per year with median barriers, for $163,000 per life saved; 11 lives per year with median strips, for $181,000 per life saved; 75 lives per year with channelled turn lanes, for $290,000 per life saved. Presumably, these programs could be expanded, with an overall cost of something like $150,000 per life saved.

It is estimated that high school courses in driver education avert about 6,000 fatalities per year and cost $180,000 per life saved, even if we include a $100 payment to each student for his or her time and trouble. Yet there are recent indications that programs are being cut back to save on costs.

Before leaving the area of traffic safety, it should be pointed out that there are 40 serious injuries for every fatality in traffic accidents. The measures we have discussed would reduce the former as well as the latter. We have therefore erred on the high side in charging all of the costs to lifesaving; the costs per life saved are lower than those given in the above discussion.

We have seen that some of our governmental agencies are passing up opportunities for saving lives at costs below $50,000, and they are rarely willing to spend over $200,000. But this does not apply to the Environmental Protection Agency (EPA) in protecting people from pollution. In its regulation dealing with air pollution control equipment for coal-burning power plants, it frequently requires installation of sulfur scrubbers, which corresponds to spending an average of $1 million per life saved. Moreover, air pollution usually affects old people with perhaps an average of 7 years of remaining life expectancy, whereas automobile accidents generally kill young people whose life expectancy averages 40 years.

Where radiation is involved, the EPA hastens to go much further. Radium is a naturally occurring element that is found in all natural waters. This has always been so, and it always will be so. However, the EPA is now requiring that in cases where radium content is abnormally high in drinking water, special measures be taken to remove some of it. This, it estimates, corresponds to spending $5 million per life saved.

Nonetheless, the EPA is not alone in being willing to spend heavily from the public purse to reduce radiation exposure. In 1972, the Office of Management and Budget recommended that nuclear reactor safety systems be installed where they can save a life for every $8 million spent. The NRC requires a $4 million expenditure per life saved in controlling normal emission of radioactivity from nuclear power plants (see Chapter 8 Appendix). But the NRC has special rules for special substances—regulations on emissions of radioactive iodine correspond to spending $100 million per life saved.

In Chapter 12 we will see that our radioactive waste management programs are spending hundreds of millions of dollars per life saved. But probably the least cost-effective spending has been on reactor safety. In Chapter 9 we will describe how the program of the NRC for improving reactor safety has increased the cost of a nuclear power plant over and above inflation by $2 billion. This program was undertaken following release of NRC's Reactor Safety Study, which estimated that over its operating lifetime, a reactor will cause an *average* of 0.8 deaths. Presumably, the $2 billion was being spent to avert these 0.8 deaths, which, by NRC's own reckoning, corresponds to $2.5 billion per life saved.

From the above discussion one gets the impression that the American public is willing to go to extremes in spending money for protection against radiation. But then there is the case of radon in our homes. Government policy here is to provide information and guidance to help citizens to protect themselves from radon, utilizing the services of private industry. The cost to the citizen for implementing this protection is about $25,000 per life saved.[12] A great deal of publicity has been given to this problem, but the public has shown little interest. Only about 2% of Americans have taken even the first step in this process of measuring the radon level in their homes, which costs about $12.

A typical attitude seems to be that we should force others to spend huge sums of money to avoid harming us, but we are not willing to spend even tiny amounts of money to protect ourselves from the same type of harm. This is an extreme manifestation of the dream of obtaining a free lunch. Somehow we haven't learned that there is no such thing, and as in other situations, we end up paying for it. The costs we inflict on utilities to provide us with super-super safety is charged back to us in our electricity bills, or in the price we pay for goods and services that require electricity for their production.

If we cut through the childish notions, we see here a truly horrible human tragedy. The $2.5 billion we spend to save a single life in making nuclear power safer could save many thousands of lives if spent on radon programs, cancer screening, or transportation safety. That means that many thousands of people are dying unnecessarily every year because we are spending this money in the wrong way.

Clearly, we have here a highly irrational situation. How did it come about? It's easy to find out. Just ask the government officials who make decisions of this type about safety requirements on nuclear plants. It turns out that at least some of these people understand the problems we are discussing. But they are powerless to follow the rational course.

The reason is that the first priority of a government official is to be responsive to public concern. That is the way a democracy operates, and that is the way we want it to operate. Anyone with a high position in government must be responsive to public concern or that person will not remain in office. Our problem, then, is *not* one of irrational behavior by government officials. It is rather a problem of misplaced public concern. The public has been very poorly educated on the hazards of radiation and of nuclear power.

Actually, the tragedy caused by this misplaced public concern is even deeper than we have described. The cost of the added safety measures on a nuclear plant, designed to save an average of less than one life, has made a coal-burning plant somewhat cheaper, and utilities are required to choose the cheapest alternative. They have therefore ordered coal-burning plants instead of nuclear. But, as discussed previously, a coal-burning plant causes an estimated 3,000 deaths over its operating lifetime. Every time a coal-burning power plant is built instead of a nuclear plant, about 3,000 people are condemned to an early death—all in an attempt to save one life.

In addition, there are indirect health costs. The increased price of power plants represents loss of wealth, and we have shown that wealth creates health. This lost wealth translates into additional deaths.

Many economists believe that a large part of the reason for America's economic success has been low-cost energy. Historically, energy in general and electricity in particular have been considerably cheaper here than in other countries. But, as a result of the cost increases under discussion here, that situation is changing. If we don't get our nuclear power program back on track, electricity will soon be much more expensive in the United States than in Western Europe or Japan. This could easily have serious effects on our standard of living and, more importantly, on our unemployment problems. It is surely not difficult to believe that this loss of our competitive advantage could result in a 1% increase in unemployment, which is estimated to cause 33,000 deaths per year.

The failure of the American public to understand and quantify risk must rate as one of the most serious and tragic problems for our nation. This chapter represents my attempt to contribute to its resolution.

Chapter 9 / COSTS OF NUCLEAR POWER PLANTS — WHAT WENT WRONG

No nuclear power plants in the United States ordered since 1974 will be completed, and many dozens of partially constructed plants have been abandoned. What cut off the growth of nuclear power so suddenly and so completely? The direct cause is not fear of reactor accidents, or of radioactive materials released into the environment, or of radioactive waste. It is rather that costs have escalated wildly, making nuclear plants too expensive to build. State commissions that regulate them require that utilities provide electric power to their customers at the lowest possible price. In the early 1970s this goal was achieved through the use of nuclear power plants. However, at the cost of recently completed plants, analyses indicate that it is cheaper to generate electricity by burning coal. Here we will attempt to understand how this switch occurred. It will serve as background for the next chapter, which presents the solution to these problems.

Several large nuclear power plants were completed in the early 1970s at a typical cost of $170 million, whereas plants of the same size completed in

1983 cost an average of $1.7 billion, a 10-fold increase. Some plants completed in the late 1980s have cost as much as $5 billion, 30 times what they cost 15 years earlier. Inflation, of course, has played a role, but the consumer price index increased only by a factor of 2.2 between 1973 and 1983, and by just 18% from 1983 to 1988. What caused the remaining large increase? Ask the opponents of nuclear power and they will recite a succession of horror stories, many of them true, about mistakes, inefficiency, sloppiness, and ineptitude. They will create the impression that people who build nuclear plants are a bunch of bungling incompetents. The only thing they won't explain is how these same "bungling incompetents" managed to build nuclear power plants so efficiently, so rapidly, and so inexpensively in the early 1970s.

For example, Commonwealth Edison, the utility serving the Chicago area, completed its Dresden nuclear plants in 1970-71 for $146/kW, its Quad Cities plants in 1973 for $164/kW, and its Zion plants in 1973-74 for $280/kW. But its LaSalle nuclear plants completed in 1982-84 cost $1,160/kW, and its Byron and Braidwood plants completed in 1985-87 cost $1880/kW — a 13-fold increase over the 17-year period. Northeast Utilities completed its Millstone 1,2, and 3 nuclear plants, respectively, for $153/kW in 1971, $487/kW in 1975, and $3,326/kW in 1986, a 22-fold increase in 15 years. Duke Power, widely considered to be one of the most efficient utilities in the nation in handling nuclear technology, finished construction on its Oconee plants in 1973-74 for $181/kW, on its McGuire plants in 1981-84 for $848/kW, and on its Catauba plants in 1985-87 for $1,703/kW, a nearly 10-fold increase in 14 years. Philadelphia Electric Company completed its two Peach Bottom plants in 1974 at an average cost of $382 million, but the second of its two Limerick plants, completed in 1988, cost $2.9 billion — 7.6 times as much. A long list of such price escalations could be quoted, and there are no exceptions. Clearly, something other than incompetence is involved. Let's try to understand what went wrong.

UNDERSTANDING CONSTRUCTION COSTS[1]

The Philadelphia office of United Engineers and Constructors (hereafter we call it "United Engineers"), under contract with the U.S. Department of Energy, makes frequent estimates of the cost of building a nuclear power plant at the current price of labor and materials. This is called the EEDB (energy economic data base), and its increase with time is plotted in Fig. 1. Circles are estimates based on the median experience (M.E.) for all plants

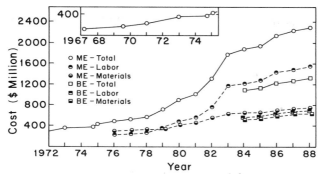

Fig. 1. The EEDB cost of a 1,000,000-kW nuclear power plant as estimated by United Engineers in various years. M.E. is median experience; B.E. is best experience; Total is labor plus materials (see text for explanation). These costs do not include escalation or interest on funds used during construction. The EEDB cost would be the actual cost if the plant were built in a very short time.

under construction at that time, while squares represent the best experience (B.E.), based on a small group of plants with the lowest costs. Also shown in Fig. 1 are the separate contributions of labor and materials. For the M.E. estimates, we see that in 1976, labor costs were substantially less than those of materials, while by 1988 they were more than twice the materials cost. During this 12-year period, labor costs escalated at an average rate of 18.7% compounded annually, the total cost escalated by 13.6%, and the materials cost escalated by 7.7%. Meanwhile, the national inflation rate was 5.7%, and the EEDB for coal-burning power plants escalated by 7.7% per year. For the B.E. situations, the annual escalation for nuclear plants was 8.4%.

There is little difference between B.E. and M.E. plants with regard to materials. They purchased the same items from the same suppliers for the same price. Incidentally, the equipment for generating electricity is purchased from vendors and represents only a small part of the materials cost— 24% for the nuclear steam supply system, which includes the reactor, steam generators, and pumps, and 16% for the turbine and generator. They represent only 7.4% and 5.0%, respectively, of the total EEDB cost. The rest of the cost is for concrete, brackets, braces, piping, electrical cables, structures, and installation.

While there is little difference in materials cost, we see from Fig. 1 that the difference in labor costs between M.E. and B.E. plants is spectacular. The comparison between these is broken down in Table 1. We see that about half of the labor costs are for professionals. It is in the area of professional

TABLE 1
BREAKDOWN OF LABOR COSTS FOR NUCLEAR POWER PLANTS
AND COAL BURNING-PLANTS FROM THE 1987 EEDB[a]

Type of labor	Median experience plant	Best experience plant	Median/ best	Coal burning
Structural craft	1.5	0.91	1.6	0.76
Mechanical craft	2.1	1.0	2.1	1.8
Electrical craft	0.80	0.48	1.7	0.52
Construction services (indirect costs)	1.7	0.86	2.0	0.38
Engineering	4.1	1.7	2.30	.56
Field supervision	3.2	0.65	4.9	0.50
Other professional	0.58	0.27	2.1	0.06
Insurance taxes	1.15	0.65	1.8	0.65
Total	15.2	6.6	2.3	5.2

Source: Ref. 1.
[a]Figures are in hundreds of 1987 dollars per kilowatt of plant capacity.

labor, such as design, construction, and quality control engineers, that the difference between B.E. and M.E. projects is greatest. It is also for professional labor that the escalation has been largest — in 1978 it represented only 38% of total labor costs versus 52% in 1987. However, essentially all labor costs are about twice as high for M.E. as for B.E. projects. The reasons for these labor cost problems will be discussed later in this chapter in the section on "Regulatory Turbulence."

The total cost of a power plant is defined as the total amount of money spent up to the time it goes into commercial operation. In addition to the cost of labor and materials which are represented by the EEDB we have been discussing up to this point, there are two other very important factors involved:

1. *The cost escalation factor (ESC), which takes into account the inflation of costs with time after project initiation.* Inflation for construction projects has been about 2% per year higher than general inflation as represented by the consumer price index (CPI).[2] For example, between 1973 and 1981, the average annual price increase was 11.5% for concrete, 10.2% for turbines, and 13.7% for pipe, but only 9.5% for the CPI.[3] For each item, the ESC depends on how far in the future it must be purchased: the basic engineering, for example, will be done shortly after the project begins and hence its cost is

hardly affected by inflation. But an instrument that can be installed rapidly and is not needed until the plant is ready to operate may not be purchased for 10 years. If the assumed inflation rate is 12% per year, which was typical of the late 1970s and early 1980s, its cost will have tripled by that time $(1.12^{10} = 3)$.

2. *A factor covering the interest charges (INT) on funds used during construction* (this is closely related to what is commonly called AFUDC, allowance for funds used during construction). All money used for construction must be borrowed or obtained by some roughly equivalent procedure. Hence the interest paid on it up to the time the plant goes into operation is included in the total cost of the plant. For example, the basic engineering may involve salaries paid 12 years before the plant becomes operational. If the annual interest rate is 15%, its cost is therefore multiplied by $(1.15^{12} =)5$. Note that the interest which increases item 2 is normally a few points higher than the inflation rate that increases item 1; it is therefore advantageous to delay money outlays for as long as possible.

Items 1 and 2 depend almost exclusively on two things, the length of time required for construction, and the rate of inflation (interest rates, averaged over long time periods are closely tied to inflation). If there were no inflation, or if plants could be built very rapidly, these factors would be close to 1.0, having little impact on the cost.

The product of these two factors, ESC × INT, used in the United Engineers estimates at various project initiation dates, is plotted in Fig. 2. The number of years required for construction is given above each point. We see

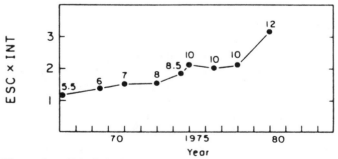

Fig. 2. The product of the inflation and interest factors. This is the factor by which the EEDB from Fig. 1 must be multiplied to obtain the total cost. The figures above the points are the estimated number of years for the project at its initiation date.

that ESC × INT was only 1.17 in 1967, when construction times were 5.5 years and the inflation rate was 4% per year. It increased to 1.45 in 1973, when construction times stretched to 8 years but inflation rates were still only 4% per year. It went up to 2.1 in 1975-1978, when construction times lengthened to 10 years and the inflation rate averaged about 7% per year, and jumped to 3.2 in 1980 when construction times reached 12 years and the inflation rate soared to 12% per year. That is, the cost of a plant started in 1980 would have been *more than triple* the EEDB cost; 69% of the final cost would have been for inflation and interest.

From this analysis we can understand two more important reasons, besides skyrocketing labor prices, that explain why costs of nuclear plants completed during the 1980s were so high: their construction times were much longer than in earlier years, and they were being built during a period of high inflation. We will now discuss the reason for the longer construction times.

REGULATORY RATCHETING

The Nuclear Regulatory Commission (NRC) and its predecessor, the Atomic Energy Commission Office of Regulation, as parts of the United States Government, must be responsive to public concern. Starting in the early 1970s, the public grew concerned about the safety of nuclear power plants: the NRC therefore responded in the only way it could, by tightening regulations and requirements for safety equipment.

Make no mistake about it, you can always improve safety by spending more money. Even with our personal automobiles, there is no end to what we can spend for safety—larger and heavier cars, blowout-proof tires, air bags, passive safety restraints, rear window wipers and defrosters, fog lights, more shock-absorbent bumpers, antilock brakes, and so on. In our homes we can spend large sums on fireproofing, sprinkler systems, and smoke alarms, to cite only the fire protection aspect of household safety. Nuclear power plants are much more complex than homes or automobiles, leaving innumerable options for spending money to improve safety. In response to escalating public concern, the NRC began implementing some of these options in the early 1970s, and quickened the pace after the Three Mile Island accident.

This process came to be known as "ratcheting." Like a ratchet wrench which is moved back and forth but always tightens and never loosens a bolt, the regulatory requirements were constantly tightened, requiring additional equipment and construction labor and materials. According to one study,[4] between the early and late 1970s, regulatory requirements increased the quantity of steel needed in a power plant of equivalent electrical output by 41%, the amount of concrete by 27%, the lineal footage of piping by 50%, and the length of electrical cable by 36%. The NRC did not withdraw requirements made in the early days on the basis of minimal experience when later experience demonstrated that they were unnecessarily stringent. Regulations were only tightened, never loosened. The ratcheting policy was consistently followed.

In its regulatory ratcheting activities, the NRC paid some attention to cost effectiveness, attempting to balance safety benefits against cost increases. However, NRC personnel privately concede that their cost estimates were very crude, and more often than not unrealistically low. Estimating costs of tasks never before undertaken is, at best, a difficult and inexact art.

In addition to increasing the quantity of materials and labor going into a plant, regulatory ratcheting increased costs by extending the time required for construction. According to the United Engineers estimates, the time from project initiation to ground breaking[5] was 16 months in 1967, 32 months in 1972, and 54 months in 1980. These are the periods needed to do initial engineering and design; to develop a safety analysis and an environmental impact analysis supported by field data; to have these analyses reviewed by the NRC staff and its Advisory Committee on Reactor Safeguards and to work out conflicts with these groups; to subject the analyzed to criticism in public hearings and to respond to that criticism (sometimes with design changes); and finally, to receive a construction permit. The time from ground breaking to operation testing was increased from 42 months in 1967, to 54 months in 1972, to 70 months in 1980.

The increase in total construction time, indicated in Fig. 2, from 7 years in 1971 to 12 years in 1980, roughly doubled the final cost of plants. In addition, the EEDB, corrected for inflation, approximately doubled during that time period. Thus, regulatory ratcheting, quite aside from the effects of inflation, *quadrupled* the cost of a nuclear power plant. What has all this bought in the way of safety? One point of view often expressed privately by those involved in design and construction is that it has bought *nothing*. A nuclear power plant is a very complex system, and adding to its complexity involves a risk in its own right. If there are more pipes, there are more ways to have pipe breaks, which are one of the most dangerous failures in reactors. With more complexity in electrical wiring, the chance for a short circuit or for an error in hook-ups increases, and there is less chance for such an error to be discovered. On the other hand, each new safety measure is aimed at reducing a particular safety shortcoming and undoubtedly does achieve that limited objective. It is difficult to determine whether or not reducing a particular safety problem improves safety more than the added complexity reduces safety.

A more practical question is whether the escalation in regulatory requirements was necessary, justified, or cost effective. The answer depends heavily on one's definition of those words. The nuclear regulators of 1967 to 1973 were quite satisfied that plants completed and licensed at that time were adequately safe, and the great majority of knowledgeable scientists agreed with them. With the exception of improvements instigated by lessons learned in the Three Mile Island accident, which increased the cost by only a few percent, there were no new technical developments indicating that more expenditures for safety were needed. In fact, the more recent developments suggested the contrary (see Chapter 6). Perhaps the most significant result of safety research in the late 1970s was finding that the emergency core cooling system works better than expected and far better than indicated by the pessimistic estimates of nuclear power opponents. Another important result was finding that radioactive iodine and other elements in a water environment behave much more favorably than had been assumed.

Clearly, the regulatory ratcheting was driven not by new scientific or technological information, but by public concern and the political pressure it generated. Changing regulations as new information becomes available is a normal process, but it would normally work both ways. The ratcheting effect, only making changes in one direction, was an abnormal aspect of regulatory practice unjustified from a scientific point of view. It was a strictly *political* phenomenon that quadrupled the cost of nuclear power plants, and thereby caused no new plants to be ordered and dozens of partially constructed plants to be abandoned.

REGULATORY TURBULENCE

We now return to the question of wildly escalating labor costs for construction of nuclear plants. They were not all directly the result of regulatory ratcheting, as may be seen from the fact that they did not occur in the "best experience" projects. Regulatory ratcheting applied to new plants about to be designed is one thing, but this ratcheting applied to plants under construction caused much more serious problems. As new regulations were issued, designs had to be modified to incorporate them. We refer to effects of these regulatory changes made during the course of construction as "regulatory turbulence," and the reason for that name will soon become evident.

As anyone who has tried to make major alterations in the design of his house while it was under construction can testify, making these changes is a very time-consuming and expensive practice, much more expensive than if they had been incorporated in the original design. In nuclear power plant construction, there were situations where the walls of a building were already in place when new regulations appeared requiring substantial amounts of new equipment to be included inside them. In some cases this proved to be nearly impossible, and in most cases it required a great deal of extra expense for engineering and repositioning of equipment, piping, and cables that had already been installed. In some cases it even required chipping out concrete that had already been poured, which is an extremely expensive proposition.

Constructors, in attempting to avoid such situations, often included features that were not required in an effort to anticipate rule changes that never materialized. This also added to the cost. There has always been a time-honored tradition in the construction industry of on-the-spot innovation to solve unanticipated problems; the object is to get things done. The supercharged regulatory environment squelched this completely, seriously hurting the morale of construction crews. For example, in the course of many design changes, miscalculations might cause two pipes to interfere with one another, or a pipe might interfere with a valve. Normally a construction supervisor would move the pipe or valve a few inches, but that became a serious rule violation. He now had to check with the engineering group at the home office, and they must feed the change into their computer programs for analyzing vibrations and resistance to earthquakes. It might take many hours for approval, and in the meanwhile, pipefitters and welders had to stand around with nothing to do.

Requiring elaborate inspections and quality control checks on every operation frequently held up progress. If an inspector needed extra time on one job, he was delayed in getting to another. Again, craft labor was forced to stand around waiting. In such situations, it sometimes pays to hire extra inspectors, who then have nothing to do most of the time. I cannot judge whether all of these new safety procedures were justifiable as safety improvements, but there was a widespread feeling among those involved in implementing them that they were not. Cynicism became rampant and morale sagged.

Changing plans in the course of construction is a confusing process that can easily lead to costly mistakes. The Diablo Canyon plant in California was ready for operation when such a mistake was discovered, necessitating many months of delay. Delaying completion of a plant typically costs more than a million dollars per day.

Since delay was so expensive, plant constructors often chose to do things that appeared to be very wasteful. Construction labor strikes had to be avoided at almost any cost. Many situations arose that justified overtime work with its extra cost. There was a well-publicized situation on Long Island where a load of pipe delivered from a manufacturer did not meet size specifications. Instead of returning it and losing precious time, the pipe was machined to specifications on site, at greatly added expense.

A major source of cost escalation in some plants was delays caused by opposition from well-organized "intervenor" groups that took advantage of hearings and legal strategies to delay construction. The Shoreham plant on Long Island was delayed[6] for 3 years by intervenors who turned the hearings for a construction permit into a circus. The intervenors included a total imposter claiming to be an expert with a Ph.D. and an M.D. There were endless days of reading aloud from newspaper and magazine articles, interminable "cross examination" with no relevance to the issuance of a construction permit, and an imaginative variety of other devices to delay the proceedings and attract media attention.

But the worst delay came after the Shoreham plant was completed. The NRC requires emergency planning exercises for evacuation of the nearby population in the event of certain types of accidents. The utility provides a system of warning horns and generally plans the logistics, but it is necessary to obtain cooperation from the local police and other civil authorities. Officials in Suffolk County, where Shoreham is located, refused to cooperate in these exercises, making it impossible to fulfill the NRC requirement. After years of delay, the NRC changed its position and ruled that in the event of an *actual* accident, the police and civil authorities would surely cooperate. It therefore finally issued an operating license. By this time the situation had become a political football, with the governor of New York deeply involved. He apparently decided that it was politically expedient to give in to the opponents of the plant. The state of New York therefore offered to "buy" the plant from the utility for $1 and dismantle it, with the utility receiving enough money from various tax savings to compensate for its construction expenditures. This means that the bill would effectively be footed by U.S. taxpayers. As of this writing, there are moves in Congress to prevent this. The ironic part of the story is that Long Island very badly needs the electricity the Shoreham plant can produce.

The Seabrook plant in New Hampshire suffered 2 years of delay[7] due to intervenor activity based on the plant's discharges of warm water (typically 80°F) into the Atlantic Ocean. Intervenors claimed it would do harm to a particular species of aquatic life which is not commercially harvested. There was nothing harmful about the water other than its warm temperatures. The utility eventually provided a large and very expensive system for piping this warm water 2½ miles out from shore before releasing it.

But again with Seabrook, the most expensive delay came after the plant was completed and ready to operate. It is located in such a way that the 5-mile radius zone requiring emergency planning extends into the state of Massachusetts. Governor Dukakis of Massachusetts, in deference to those opposed to the plant, refused to cooperate in the planning exercises. After about 3 years of delay, which added a billion dollars to the cost, in early 1990 the NRC ruled that the plant could operate without that cooperation. Governor Dukakis is appealing that decision, but the plant is now operating.

A rather different source of cost escalation is cash flow problems for utilities. When they institute a project, utilities do financial as well as technical planning. If the financial requirements greatly exceed what had been planned for, the utility often has difficulty raising the large sums of extra money needed to maintain construction schedules. It may therefore slow down or temporarily discontinue construction, which greatly escalates the final cost of the plant. For plants completed in the 1980s, this source of cost escalation was to a large extent due to regulatory turbulence, which caused the original financial planning to be so inadequate.

In summary, there is a long list of reasons why the costs of these nuclear plants were higher than those estimated at the time the projects were initiated. Nearly all of these reasons, other than unexpectedly high-inflation rates, were closely linked to regulatory ratcheting and the turbulence it created.

But what about the "best experience" plants that avoided these horrendous cost escalations. For that matter there are many plants for which the costs were much *higher* than indicated by "median experience" data. Nuclear plant costs vary by large factors. Almost every nuclear power plant built in the United States has been custom designed. This is due to the fact that, when they were designed, nuclear power was a young and vibrant industry in which technical improvements were frequently made. Varied responses to regulatory ratcheting also caused big differences between plants. The variations in cost from one plant to another have many explanations in addition to difference in design. Labor costs and labor productivity vary from one part of the country to another. Some constructors adjusted better than others to regulatory ratcheting; some maintained very close contact with the NRC and were able to anticipate new regulations, while others tended to wait for public announcements. Several different designs were used for containment buildings, and reactors that happened to have small containments had much more difficulty fitting in extra equipment required by new regulations. Some plants were delayed by intervenors, while others were not. Some had construction delays due to cash flow problems of the utilities. Plants nearing completion at the time of the Three Mile Island accident were delayed up to 2 years while the NRC was busy absorbing the lessons learned from that accident and deciding how to react to them.

Perhaps the most important cause of cost variations was the human factor. Some supervisors and designers adapt better than others to a turbulent situation. Some, considering it to be a very interesting challenge, developed ingenious ways of handling it, while others were turned off by it and solved problems unimaginatively by lavish spending of money. Some made expensive mistakes, while others were careful enough to avoid them. Some were so overwhelmed by the innumerable regulations, codes, standards, quality control audits, formal procedures for making design changes, and general red tape that they became ineffective, while others kept these problems in proper perspective and used their energies in a productive way. In some cases, people were able to cope with turbulence, but in most cases the regulatory ratcheting and the turbulence it caused exacted a terrible toll.

As a result, the average cost of nuclear electricity in the United States is now somewhat higher than that of electricity from coal burning. This represents a reversal of the situation in the 1970s and early 1980s, when nuclear energy provided the cheapest electricity. It is also the opposite of the situation in most other countries where electricity from nuclear energy is the least costly available alternative.

THE FUTURE

Regulatory ratcheting is really the political expression of difficulties with public acceptance. In an open society such as ours, public acceptance, or at least nonrejection, is a vital requirement for the success of a technology. Without it, havoc rules.

It is clear to the involved scientists that the rejection of nuclear power by the American public was due to a myriad of misunderstandings. We struggled mightily to correct these misunderstandings, but we did not succeed.

By the mid-1980s the battle was over. Groups that had grown and flourished through opposition to nuclear power went looking for other projects and soon found them. Many of them learned to distinguish between trivial problems and serious ones like global warming and air pollution.

Some of them have even made statements recognizing that nuclear power is a solution to some of those problems.

The regulatory ratcheting, of course, has not been reversed. But the nuclear industry is now developing new reactor designs that avoid most of the problems this regulatory ratcheting has brought. It is relatively easy to accommodate regulations in the initial design stages. Moreover, the new designs go far beyond the safety goals that drove the regulatory ratcheting. The nuclear industry absorbed the message that the public wants super-super safety, and they are prepared to provide it. The next chapter describes how this will be done.

But what about costs? It is useless to develop new plant designs if they will be too expensive for utilities to purchase. In fact, they must provide electricity at a substantially lower cost than that generated by coal burning. Nuclear power has an inherent disadvantage in this competition because most of its cost lies in plant construction, which the utility must pay for up front, while much of the cost of electricity from coal burning comes from buying fuel as the plant operates. Utilities have had no problem in obtaining approval from public utility commissions for charging fuel costs directly to customers in their rates. The subject of total cost of electricity will be covered in the next chapter. We will see there how recently completed nuclear power plants fail to compete with coal-burning plants, but how this situation will be remedied in the new reactor designs.

In this chapter we have pointed out lots of problems, and perhaps given the reader a gloomy outlook about the future of nuclear power in the United States. The next chapter will present the solution to these problems, and explain why the future looks bright from the standpoint of cost as well as safety.

Chapter 10 / THE NEXT GENERATION OF NUCLEAR POWER PLANTS

The nuclear power plants in service today were conceptually designed and developed during the 1960s. At that time, it was deemed necessary to achieve maximum efficiency and minimum cost in order to compete successfully with coal- or oil-burning plants. The latter were priced at 15% of their present cost and used fuel that was very cheap by current standards. In order to maximize efficiencies in the nuclear plants, temperatures, pressures, and power densities were pushed up to their highest practical limits. Safety features were exemplary for that era, and even for current safety practices in other industries. But they were not up to present-day demands for super-super safety in the nuclear industry.

As the public became more concerned with nuclear safety, the Nuclear Regulatory Commission required that new safety equipment and procedures be added on, in the process discussed in Chapter 9 as "regulatory ratcheting." The amount of labor and materials for these add-ons exceeded that for the plant as originally conceived. With this added complexity, the plants became difficult and expensive to construct, operate, and maintain. Moreover, the level of safety was still limited by the original conceptual design.

THE NEW DESIGN PHILOSOPHY[1,2,3]

By the early 1980s it became apparent that a new conceptual design of nuclear reactors was called for. The cost of electricity from coal- and oil-burning plants had escalated to the point where their competition did not require maximum efficiency from nuclear plants. Furthermore, the added efficiency achieved by pushing temperatures, pressures, and power densities to their limits was overshadowed by the efficiency lost due to shutdowns when these limits were exceeded. But above all it would be much easier to satisfy the public's demand for super-super safety by starting over with a new conceptual design than by using myriads of add-ons to a design originally targeted on rather different goals.

In the mid-1980s, several reactor vendors undertook these new designs. Let us consider some of the thinking that served as their basis.

In attempting to obtain maximum performance per unit of cost, designers nearly always find it advantageous to build plants with higher power output. This is the widely applicable principle of "economies of scale." For example, it is the reason why, before fuel costs became an issue, American automobiles were large—in stamping out an auto body, machining an engine cylinder, or in any other such operation, it doesn't cost much more to make it large than to make it small. It thus costs substantially less to build and operate a 1,200,000-kW plant than two 600,000-kW plants of the same basic design. In line with the old goals, plants were therefore built in the former size range, and constantly grew larger. This trend still dominates in foreign countries.

However, if safety is the primary goal, as it is in the United States today, it is much easier to assure that adequate cooling will be available after shutdown if there is only half as much heat to dissipate. In fact, in a 1,200,000-kW reactor, cooling requires elaborate pumps, while in a 600,000-kW reactor it can be handled by simple gravity flow with natural convection—cool water enters the bottom of the reactor, which heats it, causing it to rise because warm water is lighter. This process sets up a natural circulation driven only by gravity. Unlike pumps which can fail and are driven by electric power which may not always be available, gravity never stops working. That makes the 600,000-kW reactor inherently safer. This is called "passive stability," since no active measures by operators or by mechanical or electrical control systems are required. The operator could shut off the electric power and go home without any harm coming to the reactor.

Natural circulation can also be used to protect the containment from breaking open due to excess pressure. In present-day power plants, active cooling using water pumps is necessary to control the pressure. But with the smaller reactor, there is less energy to dissipate, making natural circulation a viable alternative.

Of course the safety of the larger reactor system can always be improved by adding extra pumps and extra diesel-driven generators to provide power for them in the event of failures, but this drives up the cost. Moreover, this extra equipment must be periodically tested, maintained, and repaired. The control systems that start the equipment when it is needed, and open and close the necessary valves, must be maintained and frequently tested. As a result of problems like these, a super-super safe 1,200,000-kW reactor might cost almost as much as two super-super safe 600,000-kW reactors. Moreover, the 1,200,000-kW reactor is still not as safe because human failures can enter the picture—for instance, a person can push the wrong button in starting a pump or opening a valve, or the person who installed an automatic system for performing these functions can make a mistake in wiring. In addition, with more pipes entering the reactor there are more opportunities for a pipe to break, causing a loss-of-coolant accident (LOCA).

We have seen that the time required for construction can be a very important ingredient in determining costs, and it obviously takes longer to construct a larger plant. This was not an important factor in the early 1970s, when large plants could be built in 4 or 5 years, as they still are today in France and Japan. But with the very elaborate and time-consuming practices now required in the United States, construction times have become an important factor in determining the relative costs of large and small reactor plants. For example, with elaborate tests and documentation for these tests required for every weld, reducing the number of required welds results in substantial time saving. Since they have fewer pumps and valves and less piping, smaller reactor plants require fewer welds, which reduces the time needed for constructing them.

Still another consideration is that many more parts of a smaller reactor than of a larger one can be produced and assembled in a factory. Operations are much more efficient in a factory than at a construction site because permanently installed equipment and long-term employees can be used to produce many copies of the same item for many different power plants. Use of transient construction workers and portable equipment to do a wide variety of jobs just once introduces all sorts of inefficiencies.

When all of these matters are taken into account, it turns out that if super-super safety is a primary concern, two 600,000-kW reactors may even be cheaper and are certainly considerably safer than a single 1,200,000-kW reactor.

In France and England, one government-owned utility provides electricity for the whole country. But in the United States there are hundreds of utilities, and they are generally rather small. Consider a typical American situation: a particular utility's service load is growing by 100,000-kW per year, and signs indicate that it will continue to grow at that rate for the next few years. A 600,000-kW plant might be an attractive investment to take care of this growth, but a 1,200,000-kW plant would be much more expensive and would provide a lot more power than needed for a long time. Moreover, it is hard to predict how long the growth will continue; if it should fall off, the utility may never need the full 1,200,000-kW. Thus a 600,000-kW size is better suited to the needs of an American utility. This is not the case for the huge French and British national utilities, which generally have a load growth of several million kilowatts per year.

Since the 600,000-kW size range is optimum for U.S. utilities, reactor vendors are competing to provide them. In fact, not only are U.S. reactor vendors developing them, but foreign vendors who build larger reactors for their own countries are developing 600,000-kW reactors to sell to American utilities. Coal- and oil-burning power plants of this size have always been available and are the most widely used plants in the United States.

Another aspect of the new design philosophy is to favor a larger tolerance to variation in operating conditions over optimizing efficiency. This represents an abandonment of the controlling principle of reactor design in the 1960s. For example, optimum plant efficiency requires heating the water to the highest possible temperature, but increased temperature accelerates corrosion of steam generator tubes. In the new design, water temperatures are reduced from 615°F to 600°F, and, in addition, a new stainless steel alloy that is more durable under exposure to high-temperature water is used for steam generator tubes. As another example, the reactor must be shut down for safety reasons if the power density—the energy per second produced by each foot of fuel assembly—exceeds a predetermined limit. Present reactors operate at 5.5 kW per foot, but in the new reactors this is reduced to 4.0 kW per foot. The change reduces efficiency, but it means fewer shutdowns due to excessive power density.

Another change in the new design is to favor simplicity over complexity. Adding complex equipment and operations may improve the efficiency of a reactor, but it also introduces more possibilities for equipment failures and operator mistakes. In earlier reactors, living with these problems was deemed to be a worthwhile sacrifice, but the new design philosophy is that it is not.

In addition to changes in design philosophy, there have been many lessons learned from experience that can be incorporated into the new reactors. These include systems for adding small quantities of various chemicals to the water to reduce corrosion problems, employing different materials for certain applications, and using new methods for construction. The principal lesson learned from the Three Mile Island accident is the significance of making important information easily available to the operator. The revolution in computer and telecommunications technology over the last two decades offers bountiful opportunities for improvements here in revamping the control room and instrumentation systems, and providing graphic displays, diagnostic aids, and expert systems to handle a variety of situations. Because of the passive stability features, there are few situations which require the operator to take emergency action. This also makes the reactor easier to operate.

LICENSING REFORM

Licensing of reactors has become a major problem for U.S. utilities. They must obtain a construction permit to start building the plant, and when the plant is nearing completion they must apply separately for an operating license. These are long, tedious, expensive processes that sometimes cause disastrous delays. In Chapter 9 we mentioned that the Shoreham and Seabrook plants encountered delays of several years, costing over a billion dollars for each, while the utilities struggled against political opposition to obtain an operating license. To avoid this sort of fiasco in future plants, the NRC is developing new rules under which both a construction permit and an operating license are obtained in a single procedure before construction is started. When the plant is completed, the utility need only show that it was constructed in accordance with the plans originally approved. It remains to be seen whether Congress will go along with these new rules. The interaction between the NRC and Congress is a matter of considerable complexity, and utilities are understandably wary of being caught in the middle.

Even more important in the licensing problem is the matter of standardization. Almost every U.S. reactor is a custom design and custom built. In the late 1960s and early 1970s, this seemed appropriate because lots of new ideas were available for trial, and designs were changing rapidly on the basis of earlier experiences. By the mid-1970s, things were settling down, and American reactor vendors began developing standard designs, but that was just when the orders for new reactors stopped. In France, however, reactor construction was just beginning to burgeon at that time, and the French government-owned vendor adopted the standardized design approach. This proved to be very efficient and successful, resulting in high-quality plants built quickly and cheaply. Construction crews, after having built several identical plants, knew exactly what to do, where to look for trouble spots, and how to correct them. Once a new reactor went into service, the operating crews were experienced from the start, since most of their members had done the same job before in identical plants.

For the United States, with its current tumultuous licensing procedures, standardization would appear to be a panacea. A reactor vendor would go through the laborious and expensive process of getting its standard design certified by the NRC. But once this was accomplished, a utility would need only prove that its site is suitable, a very much simpler procedure. Once this was done, they would have clear sailing with no delay before going into commercial operation. Since all equipment and procedures would be standardized, there would be much less that could go wrong.

THE NEXT GENERATION[4]

In the mid-1980s, Electric Power Research Institute (EPRI), the private research arm of the electric utility industry, undertook to set design goals for new reactors that make use of currently available technology and incorporate the philosophy expounded above. That project developed the following specifications:[1,5]

1. Reducing the probability of a core damage accident by at least a factor of 10 compared to reactors presently in commercial operation—a much larger factor is now anticipated by those designing and evaluating the new reactors.
2. Developing a much simpler design with larger margins for safety.
3. Achieving an operating lifetime of 60 years, and an availability for producing power at least 85% of the time (versus the present average of 70% for both nuclear and coal-burning plants).
4. Providing electricity for lower cost than coal-burning plants. (Present indications are that it will be at least 15% lower.)
5. Completing construction in 3 to 5 years (versus more than 12 years in recently completed nuclear plants).

6. Producing a design standardized at a high-quality level with licensing assured.

A mammoth effort both in the United States and abroad is now underway to develop reactors to these specifications. The principal designs that have emerged include a pressurized water reactor called AP-600 (for advanced passive 600,000 kW), developed by Westinghouse and partner companies[6]; a boiling water reactor called SBWR (where S stands for simpler, smaller, safer), developed by General Electric, Bechtel, and the Massachusetts Institute of Technology; a liquid-metal-cooled reactor called PRISM (for power reactor inherently safe module), developed by General Electric in collaboration with Argonne National Laboratory; and a gas-cooled reactor called MHTGR (modular high-temperature gas-cooled reactor), developed by GA Technologies plus a consortium of other companies. Utilities from France, Italy, the Netherlands, Japan, Korea, and Taiwan have also participated with these companies in developing the designs.

In addition, large programs for developing advanced reactor designs have emerged in France, Britain, Germany, and Japan, although these have been on larger reactors without the passive stability features. A Swedish effort[7,8] is developing the PIUS (process inherent ultimate safety) reactor, a 640,000-kW PWR with essentially no dependence on active equipment.

In the United States, the AP-600 and the SBWR are expected to pass the licensing requirements and be ready for commercial orders by 1995, but the PRISM and MHTGR will not reach that stage until after the turn of the century.

For the interim of the next few years, American reactor vendors have developed improved versions of the reactors now in service, with several changes targeted on reducing complexity and improving safety. They also expect to have a long-term market for these reactors, called the APWR and ABWR (A for advanced), in foreign countries. One of them has already been ordered for construction in Japan. The modifications reduce the probability of a severe accident tenfold, have increased margins for error in key areas, and give operators more time to respond to emergency situations. These reactors should be certified for licensing in the United States by 1992.

Let us examine the salient features of some of the next-generation reactors.

The AP-600[6]

The AP-600 obtains its emergency cooling from huge water tanks mounted above the reactor. Some of these are pressurized with nitrogen gas, allowing them to inject water even if the reactor remains at high pressure, as it may in some accident scenarios. In most cases, neither electric power nor operator action are needed to start injection. For example, if the pressure in the reactor falls due to a break in the system, the valves connecting some of the tanks to the reactor are automatically pushed open by the fact that the pressure on the tank side is higher than on the reactor side. Actually, present reactors have similar systems. The "accumulators" mentioned in Chapter 6 operate on the same principle, but with enough water for only about 15 minutes, as compared with several hours in the AP-600.

One of the large tanks above the reactor serves as a place to deposit heat after shutdown. It is connected to the reactor by two pipes, one leading to the bottom of the reactor vessel and one to the top. As water in the vessel is heated, it automatically rises in the upper pipe and is replaced by cool water from the lower pipe, establishing a natural circulation which transfers heat from the reactor into the water tank. This is analogous to air heated by a radiator in a house rising (because it is lighter) and spreading through the room to transfer the heat from the radiator to all the air in the room.

Water tanks pressurized with nitrogen gas also provide sprays to cool the atmosphere inside the containment and remove some of the volatile radioactive materials from the air in the event of an accident. Again, no pumps are needed.

The steel containment shell is cooled by air circulating between it and the concrete walls, again by gravity-induced convection. In addition there is water draining by gravity onto the containment shell, though the air circulation alone is sufficient to provide the necessary cooling. Air circulation keeps the pressure inside below 40 pounds per square inch in the worst accident scenarios.

Probabilistic risk analyses yield estimates that a core damage accident can be expected only once in 800,000 years of reactor operation, and that there is less than a 1% chance that this will be followed by failure of the containment. This makes the AP-600 a thousand times safer than the current generation of reactors. It is also much simpler, reducing the number of valves by 60%, large pumps by 50%, piping by 60%, heat exchangers by 50%, ducting by 35%, and control cables by 80%. The volume of buildings required to have a very high degree of earthquake resistance is thereby reduced by 60%. It is estimated that the plant can be constructed in 3 to 4 years. All of these factors contribute to reducing the cost.

Other Next-Generation Reactors[4,9]

In a pressurized water reactor (PWR) the water is only heated, and it can therefore flow directly from the inlet to the outlet, passing through the core only once. But in a boiling water reactor (BWR), all the water cannot suddenly turn into steam as it passes through the reactor core; boiling occurs, releasing bubbles of steam which rise to the top of the reactor vessel, but most of the water remains in liquid form. After passing through the core it must therefore be recirculated again and again with a fraction converted to steam on each pass. This recirculation is now done with pumps outside the reactor, but in the SBWR it will be done by natural circulation entirely inside the reactor.

The modification of the emergency core cooling system in the SBWR to achieve passive operation by natural convection is similar in principle to what is described above for the AP-600 PWR. In addition, new stainless steel alloys, advanced welding techniques, and improved water chemistry are used to eliminate stress corrosion cracking experienced in present BWRs. Reductions in the amount of piping, valves, and the like, comparable to those in the AP-600 are achieved to reduce complexity and cost. The SBWR should also be ready for orders by utilities in the mid-1990s.

The PRISM uses liquid sodium metal rather than water to transfer the heat. This has the advantage that the system does not operate under high pressure and therefore cannot burst open, which is a plus for safety. The disadvantage is that sodium burns in air or in water, releasing large quantities of radioactive materials if it gets out of the reactor. But there is now a great deal of technology and experience with safe handling of sodium.

The PRISM would be built in small modules producing 155,000 kW each. If the electric power should fail, the operators could leave without harm to the reactor. Its heat would be radiated to the containment shell, which would be cooled by natural air circulation. The reactor and even the containment vessel can be fabricated in a factory and shipped by railroad, making installation at the site very quick and easy.

Part of the reason for development of the PRISM is that it provides advantages in fuel fabrication, reprocessing of spent fuel, waste management, and protection against theft of plutonium. This design concept has a great deal of potential from many points of view, but a great deal of development and testing will be required before it can be evaluated in its totality.

The MHTGR is also modular and even smaller at 135,000 kW than the PRISM. The heat is transferred out of the reactor with helium gas, which allows use of much higher temperatures and thereby gives higher efficiency. The fuel consists of small particles of uranium plus thorium (which is converted to fissile uranium in the reactor), sealed inside very high melting point materials and encased in huge quantities of graphite, which can absorb tremendous amounts of heat. At any temperature above 1000°F, it radiates this heat to the surroundings much faster than it can be generated, leaving no possibility for the fuel to even approach the 4000°F temperature where it might be damaged. It is therefore believed that this reactor will not even require a containment. The modular character allows factory fabrication and quick, cheap construction.

Since they incorporate so many innovative features, prototypes will have to be constructed and tested before either the PRISM or the MHTGR can become candidates for purchase by utilities. This is not true of the AP-600 or the SBWR, because they represent relatively minor departures from well-established technologies. We can expect orders for these plants in the mid-1990s.

COST PER KILOWATT-HOUR

Quantities of electrical energy are expressed in kilowatt-hours. For example, our homes are billed for electricity use at a certain cost per kilowatt-hour, currently ranging from 8 cents to 12 cents. However, that billing includes distribution costs—bringing the electricity from the power plant into our homes—and these costs are substantial. Here we consider only the cost per kilowatt-hour as it leaves the power plant.

The history of declining electricity costs through the first three-quarters of this century is one of the great success stories of technology. In the 1930s, electricity was an expensive commodity. You never left a room without turning off the light, and keeping a refrigerator door open longer than necessary was a serious transgression. But as time passed, the cost per kW-h dropped steadily, even while the price of everything else was doubling and redoubling. By the early 1970s, it had dropped to 1.5 cents per kilowatt-hour, and electricity was being used profligately in all sorts of applications. With the energy crisis of 1974 raising the price of fuel, and the general inflation following it, the cost of electric power began to rise. By 1981, the average costs per kilowatt-hour were 2.7 cents for nuclear, 3.2 cents for coal, and 6.9 cents for oil.[10] The low average price for nuclear power was due to the low-cost plants completed before 1975, for which the average cost was 2.2 cents versus 3.5 cents for those completed after 1975. As more new nuclear plants were completed, the average cost of nuclear power increased. A survey by the Atomic Industrial Forum found the costs in 1985 to be 4.3 cents for nuclear, 3.4 cents for coal, and 7.3 cents for oil.

The U.S. Energy Information Agency does not try to calculate averages but rather reports on the range of prices for all large plants. In 1984 it gave 1.52 to 8.17 cents for nuclear versus 1.86 to 6.41 cents for coal, and in 1987 it was 1.68 to 8.50 cents for nuclear versus 1.82 to 7.98 cents for coal.[11] These very wide ranges are indicative of how little meaning there is to averages over plants now in operation. For example, they depend very much on the mix of old and new plants in use, since there is no inflation factor applied to the original cost of the plant.

More germane to our interests here are the estimated costs of electricity from plants to be constructed in the future, since this is what is important in a utility's choice of a new plant. In Table 1, we break down the cost per kilowatt-hour for various candidates.[12] The cost of operating and maintaining a facility, and of providing it with fuel are easy to understand. The contribution of the capital used in constructing the plant is calculated as the amount of money per kilowatt-hour needed to purchase an annuity which will pay off all capital costs and interest before the end of the facility's useful life. The decommissioning charge is the amount of money per kilowatt-hour that must be set aside so that, including the interest it has earned, there will be enough money to pay for decommissioning the plant when it is retired. There are many uncertainties in these calculations, but both the U.S. Government and the utilities employ groups of experts who have developed standard procedures that are normally applied and widely accepted. A utility gambles billions of dollars on decisions based on these analyses, so I would think they do everything possible to make them as reliable as possible.

The first two entries in Table 1 are data for present-generation reactors based on median experience and best experience as defined and used in Chapter 9. The APWR (advanced pressurized water reactor) refers to present-type reactors with various upgrades to improve safety, a standardized design, and the streamlined licensing procedures discussed above. The

TABLE 1
COST PER KILOWATT-HOUR FOR VARIOUS TYPES OF POWER PLANTS

Type	Cost[a]				
	Capital	Operation and maintenance	Fuel	Decommissioning	Total
Median experience	56	13.0	7.2	0.5	77
Best experience	28	9.1	6.4	0.5	44
APWR	22	9.1	6.4	0.5	38
AP-600 (2)	22	10.4	6.4	0.7	40
Coal (2)	21	5.9	21.0	0.1	48

[a]Figures are in mills (0.1 cents) of 1987 dollars, for plants going into operation in the year 2000.

AP-600 represents the new generation of PWRs and BWRs that fully utilize passive safety features as we have described. Both the AP-600 and the coal-burning plant are of half the capacity of the present generation reactors and the APWR—600,000 kW versus 1,200,000 kW—and Table 1 therefore considers two of these plants at the same site in drawing comparisons. Note that having two plants rather than one increases the operating, maintenance, and decommissioning costs and makes electricity from the AP-600 slightly more expensive than from the APWR. That is part of the price we pay for super-super safety, but it is hardly significant.

The lower capital cost of the APWR than of the "best experience" present-generation reactors is based on applying lessons learned from the latter, benefits from standardization and easier licensing, and freedom from regulatory turbulence. Actually the only large difference in Table 1 is the gap between the median experience and best experience present-generation reactors. While the costs of the others are similar, the APWR is about 10 times safer, and the AP-600 is 1,000 times safer than the present-generation reactors.

The most important point in Table 1 is that electricity from the new-generation reactors is about 20% cheaper than from a coal-burning plant. The coal-burning plant is cheaper to build, operate, maintain, and decommission, but its fuel cost is more than 3 times higher.

The fact that future nuclear power plants will produce electricity at a substantially lower price than coal-burning plants is confirmed by a more elaborate study by industry analysts.[13] Their study also concludes that the cost of electricity from future nuclear plants will be about 20% cheaper than from coal-burning plants. It explores the sensitivity of this conclusion to a wide range of uncertain parameters used in the calculations, but always finds nuclear power to be less expensive.

Of course, this conclusion depends heavily on the assumptions of standardized design, streamlined licensing procedures, and no regulatory turbulence. These would reduce construction times to 5-6 years, according to the Department of Energy analysts who developed Table 1, and that fact alone would be responsible for a large fraction of the cost saving. Industry planners are hoping for even shorter construction times, especially after a few of these standardized plants have been built.

In the past, many other analyses have been reported on cost comparisons between nuclear and coal burning, always based on the assumption of reasonable (i.e., nonturbulent) construction practices. This is clearly a very practical question for a utility deciding to build a new power plant. Many utilities seek cost analyses from economics consulting firms, some utilities have their own in-house economists to make estimates, and banking organizations maintain expertise to aid in decisions on investments. From the early 1970s until the early 1980s, all of their reports found that nuclear power was the cheaper of the two. For example, the Tennessee Valley Authority (TVA) is the largest electric utility in the United States. Its profits, if any, are turned back to the U.S. Treasury. It maintained a large and active effort for many years in analyzing the relative cost advantages of nuclear versus coal-burning power plants, consistently finding that nuclear power was cheaper. The 1982 analysis by the Energy Information Administration, a branch of the U.S. Department of Energy, was the first to find that coal and nuclear were equal in cost; their previous analyses found nuclear to be cheaper. By 1982, these analyses were mostly discontinued, since it seemed unrealistic for a utility to consider building a nuclear power station, or even to hope that it could be done without regulatory turbulence. Only with the new optimism about the future of nuclear power have these analyses been resumed.

There have been claims that utilities are biased toward nuclear and against coal because they can make more profits from the former. This is hard to understand in view of the fact that many utilities have been badly hurt financially by their nuclear ventures. It would not explain the positions of nonprofit utilities like TVA or municipally owned power authorities. Most Western European power plants are built and owned by the government, with no possible profit motive in favoring the nuclear option, and of course the same is true of Eastern European nations, the Soviet Union, and China, all of whom are actively pursuing nuclear power. The last two, at least, also have large reserves of coal.

In Western Europe, where there has been no regulatory turbulence comparable to that in the United States, most nuclear power plants have been relatively cheap. A 1982 study by the European Economic Community[14] estimated that for projects started at that time, electricity from nuclear plants would be less costly than that from coal-fired plants by a factor of 1.7 in France, 1.7 in Italy, 1.3 in Belgium, and 1.4 in Germany. These results are based on only a very slow rise in the price of coal, and do not include scrubbers for removal of sulfur. Scrubbers are usually required on new coal-burning plants in the United States and add very substantially to their cost.

With safety problems hopefully behind us, and with cost considerations looking favorable, it truly seems like the United States is now ready for Nuclear Power: Act II.

Chapter 11 / HAZARDS OF HIGH-LEVEL RADIOACTIVE WASTE— THE GREAT MYTH

An important reason for the public's concern about nuclear power is an unjustifiable fear of the hazards from radioactive waste. Even people whom I know to be intelligent and knowledgeable about energy issues have told me that their principal reservation about use of nuclear power is the disposal of radioactive waste. Often called an "unsolved problem," many consider it to be the Achilles' heel of nuclear power. Several states have laws prohibiting construction of nuclear power plants until the waste disposal issue is settled. Yet ironically, there is general agreement among the scientists involved with waste management that radioactive waste disposal is a rather trivial technical problem. Having studied this problem as one of my principal research specialties over the past 15 years, I am thoroughly convinced that radioactive waste from nuclear power operations represents less of a health hazard than waste from any other large technological industry. Clearly there is a long and complex story to tell.

A FIRST PERSPECTIVE

What is this material that is so controversial? As we know from elementary physical science courses, matter can be neither created nor destroyed. When fuel is burned to liberate energy, the fuel doesn't simply disappear. It is converted into another form, which we refer to as "waste." This is true whether we burn uranium or coal or anything else. For nuclear fuels, this residue, called "high-level waste," has been the principal source of concern to the public.

As an initial perspective, it is interesting to compare nuclear waste with the analogous waste from a single large coal-burning power plant. The largest component of the coal-burning waste is carbon dioxide gas, produced at a rate of 500 pounds every second, 15 tons every minute. It is not a particularly dangerous gas, but it is the principal contributor to the "greenhouse effect" discussed at some length in Chapter 3. The other wastes from coal burning were also discussed in Chapter 3, but let's review them briefly. First and probably foremost is sulfur dioxide, the principal cause of acid rain and perhaps the main source of air pollution's health effects, released at a rate of a ton every 5 minutes. Then there are nitrogen oxides, the second leading cause of acid rain and perhaps also of air pollution. Nitrogen oxides are best known as the principal pollutant from automobiles and are the reason why cars need expensive pollution control equipment which requires them to use lead-free gasoline; a single large coal-burning plant emits as much nitrogen oxide as 200,000 automobiles. The third major coal burning waste is particulates including smoke, another important culprit in the negative health effects of air pollution. Particulates are released at a rate of several pounds per second. And next comes the ash, the solid material produced at a rate of 1,000 pounds per minute, which is left behind to cause serious environmental problems and long-term damage to our health. Coal-burning plants also emit thousands of different organic compounds, many of which are known carcinogens. Each plant releases enough of these compounds to cause two or three cancer deaths per year.[1] And then there are heavy metals like lead, cadmium, and many others that are known or suspected of causing cancer, plus a myriad of other health impacts. Finally there is uranium, thorium, and radium, radioactive wastes released from coal burning that serve as a source of radon gas. The impact of this radioactive radon gas from coal burning on the public's health far exceeds the effects of all the radioactive waste released from nuclear plants (see Chapter 12).

The waste produced from a nuclear plant is different from coal-burning wastes in two very spectacular ways. The first is in the quantities involved:

the nuclear waste is 5 million times smaller by weight and billions of times smaller by volume. The nuclear waste from one year of operation weighs about 1½ tons[2] and would occupy a volume of half a cubic yard, which means that it would fit under an ordinary card table with room to spare. Since the quantity is so small, it can be handled with a care and sophistication that is completely out of the question for the millions of tons of waste spewed out annually from our analogous coal-burning plant.

The second pronounced difference is that the nuclear wastes are radioactive, providing a health threat by the radiation they emit, whereas the principal dangers to health from coal wastes arise from their chemical activity. This does not mean that the nuclear wastes are more hazardous; on nearly any comparison basis the opposite is true. For example, if all the air pollution emitted from a coal plant in one day were inhaled by people, 1½ million people could die from it,[3] which is 10 times the number that could be killed by ingesting or inhaling the waste produced in one day by a nuclear plant.[4]

This is obviously an unrealistic comparison since there is no way in which all of either waste could get into people. A more realistic comparison might be on the basis of simple, cheap, and easy disposal techniques. For coal burning this would be to use no air pollution control measures and simply release the wastes without inhibition. This is not much worse than what we are doing now since the smoke abatement techniques, which are the principal pollution control on most coal-burning plants, contribute little to health protection. We have seen in Chapter 3 that air pollution causes about 75 deaths per year from each plant. Some other comparably serious consequences will be considered later in this chapter.

For nuclear waste, a simple, quick, and easy disposal method would be to convert the waste into a glass—a technology that is well in hand—and simply drop it into the ocean at random locations.[5] No one can claim that we don't know how to do that! With this disposal, the waste produced by one power plant in one year would eventually cause an average total of 0.6 fatalities, spread out over many millions of years, by contaminating seafood. Incidentally, this disposal technique would do no harm to ocean ecology. In fact, if all the world's electricity were produced by nuclear power and all the waste generated for the next hundred years were dumped in the ocean, the radiation dose to sea animals would never be increased by as much as 1% above its present level from natural radioactivity.

We thus see that if we compare the nuclear and coal wastes on the basis of cheap, simple, and easy disposal techniques, the coal wastes are (75 ÷ 0.6 =) 120 times more harmful to human health than nuclear waste.

This treatment ignores some of the long-term health effects of coal burning to be discussed later.

Another, and very different, way of comparing the dangers of nuclear and coal waste is on the basis of how much they are changing our exposures to toxic agents. The typical level of sulfur dioxide in the air of American cities is 10 times higher than natural levels, and the same is true for the principal nitrogen oxides.[6] For cancer-causing chemicals the ratio is much higher.[7] These are matters that might be of considerable concern in view of the fact that we really do not understand the health effects of these agents very well. For radioactive waste from a flourishing nuclear industry, on the other hand, radiation exposures would be increased by only a tiny fraction of 1% above natural levels.[8]

Another basis for comparison between the wastes from nuclear and coal burning is the "margin of safety," or how close average exposure is to the point where there is direct evidence for harm to human health. There have been several air pollution episodes in which there were hundreds of excess deaths, with sulfur dioxide levels about 1,000 micrograms per cubic meter,[1] whereas levels around 100 micrograms per cubic meter, just 10 times less, are quite common in American cities. For radiation, on the other hand, exposures expected from hypothetical problems with nuclear waste are in the range of 1 mrem or less, tens of thousands of times below those for which there is direct evidence for harm to human health.

All of these comparisons have been intended as an introduction to the subject. Now, let us look into the problem in some detail.

HIGH-LEVEL RADIOACTIVE WASTE

The fuel for a nuclear reactor is in the form of ceramic pellets of uranium oxide, each about ½ inch in diameter and ¾ inch long, lined up and sealed inside metal tubes about 12 feet long and ½ inch in diameter. About 50 of these "fuel pins" are bound together, leaving space for water to circulate between the pins, to form a "fuel assembly." A large reactor might contain 200 of these fuel assemblies arranged in a lattice, with a total of 110 tons of uranium. A typical fuel assembly remains in the reactor for about 3 years. During this time, a small percentage of the uranium nuclei are struck by neutrons, causing them to undergo fission or other nuclear reactions. These reactions transform them into other nuclei, "the high-level radioactive waste." At the end of this 3-year period, the fuel is "spent" and must be replaced by fresh fuel. The subject we examine here is disposal of this spent fuel containing the radioactive waste.

One option is to dispose of the spent fuel directly by sealing fuel assemblies into canisters and burying them. A much more rational approach from the standpoint of long-range planning is to put the fuel through a chemical reprocessing operation to separate out the uranium and plutonium which are valuable for future use as fuels, and to convert the residual material into a form suitable for burial. If we follow the first option, our uranium resources will run out some time during the next century, whereas the second option can provide all of the energy humans will need for as long as they inhabit planet Earth, without an increase in fuel costs (see Chapter 13).

Unfortunately, decisions are not made by long-range planners, but rather by business executives and politicians who base their decisions on short-term economics. Even from this standpoint, reprocessing is the more favorable option, albeit by a rather narrow margin. Consequently, reprocessing plants have been an important part of nuclear industry planning in the United States. In the late 1960s and early 1970s, three commercial plants were built, one of them was operated for a few years, and still other plants reached the advanced planning stage. However, the opposition to nuclear power in the late 1970s caused severe setbacks for the reprocessing industry, leading to its demise (see Chapter 13). Therefore, current U.S. plans are to bury the spent fuel directly.

The situation is very different in other countries. Great Britain has been reprocessing spent fuel since the 1950s and is currently constructing a large new plant. France has been operating two reprocessing plants for many years and its third plant is nearly complete. Germany has been operating a small reprocessing plant since 1971, and a much larger one is under construction with a 1996 start-up date. Japan has a small plant in operation; a much larger one, under construction, is scheduled to come on line in 1995. The Japanese also send some spent fuel to France and Great Britain for reprocessing. Belgium, Italy, Switzerland, and the Netherlands also utilize the French and British facilities. India has one reprocessing plant in operation and has recently begun construction of another. China, where nuclear power is in its infancy, is planning to operate a pilot plant by 1995 and a full-scale unit by 2010. The Soviet Union reprocesses its spent fuel, and Eastern bloc countries as well as Finland make use of the Soviet facilities. Sweden, Canada, and Spain are providing for long-term storage of spent fuel, and are therefore deferring a decision on reprocessing.

Only the United States has a full-scale program in operation leading to burial of spent fuel, although it would be very easy to change it to include reprocessing. This option will remain open for a long time, since even after burial the spent fuel will be retrievable for several decades.

While I fully understand the basis for the U.S. policy of burying spent fuel and I support it, I cannot believe that it will persist for more than 20 years or so. I therefore base the risk evaluations on the assumption that we will eventually do reprocessing. If I am wrong on this matter and spent fuel is permanently buried, the risks we will be discussing will be increased about ten-fold. Nevertheless it will be evident that this would not appreciably affect the conclusions of our discussion.

HAZARDS AND PROTECTIVE BARRIERS

Probably the most frequent question I receive about nuclear power is, "What are we going to do with the radioactive waste?" The answer is very simple: we are going to convert it into rocks and put it in the natural habitat for rocks, deep underground. The next question is usually, "How do you know that will be safe?" Answering that question is our next topic for consideration.

This discussion will be in terms of the high-level waste produced by one large power plant (1,000,000 kW) in one year. The high-level waste from such a plant is contained inside the fuel. If we are to have reprocessing, the 35 tons of spent fuel will be shipped to a chemical reprocessing plant where it will be dissolved in acid and put through chemical processes to remove 99.5% of the uranium and plutonium that are valuable as fuels for future use. The residue—1.5 tons of high-level waste—will then be incorporated into a glass in the form of perhaps 30 cylinders, each about 12 inches in diameter and 10 feet long, weighing about 1,000 pounds. There is a great deal of research on materials that may be superior to glass as a waste form, but far more is now known about glass, a material whose fabrication is simple and well developed. There is no evidence that it is not satisfactory. Many natural rocks are glasses, and they are as durable as other rocks.*

This 15 tons of waste glass, roughly one truckload, will then be shipped to a federal repository, where it will be permanently emplaced deep underground. The U.S. Department of Energy is now spending over $200 million per year in developing this waste storage technology. The program is paid for by a tax on nuclear electricity; about $6 billion was collected by the end of 1989. The estimated cost of handling and storing our one truckload of waste

---

*From our experience with tumblers, jars, and trinkets, we think of glass as fragile, but a large block of solid glass is not much more fragile than most rocks. Both can crack and chip when forces are applied, but this does not greatly affect their underground behavior since there are powerful pressures holding the pieces together.

is $5 million, which corresponds to 0.07 cents per kilowatt-hour, about 1% of the cost of electricity to the consumer.

I have often been told, "We don't want a waste dumping ground in our area." This conjures up a picture of dump trucks driving up and tilting back to allow their loads of waste to slide down into a hole in the ground. Clearly it doesn't cost $5 million for each truckload to do this, especially on a mass-handling basis. What is planned, rather, is a carefully researched and elaborately engineered emplacement deep underground in a mined cavity.

It may be of incidental interest to point out that the present $5 million price tag on storing one year's waste from one plant is several hundred times higher than the cost of the original plan as formulated in the 1960s. Those conversant with the earlier plan are still convinced that it provided adequate safety, although the added expenditures do, of course, contribute further safety. This tremendous escalation in the scope of the project is a tribute to the power of public hysteria in a democracy like ours, whether or not that hysteria is justified. It is also an application of Parkinson's Law—it was clearly acceptable to devote 1% of the cost of electricity to disposal of the waste, so the money was "available," and being available, the government has found ways to use it.

There have been many safety analyses of waste repositories,[9] and all of them agree that the principal hazard is that somehow the waste will be contacted by groundwater, dissolved, and carried by the groundwater into wells, rivers, and soil. This would cause contamination of drinking water supplies or, through pick-up by plant roots, contamination of food. It could thereby get into human stomachs.

The chance of this waste becoming suspended in air as a dust and inhaled by humans is very much less because groundwater rarely reaches the surface; moreover, we inhale only about 0.001 grams of dust per day (and 95% of this is filtered out by hairs in the nose, pharynx, trachea, and bronchi and removed by mucous flow), whereas we eat over 1,000 grams of food per day. On an average, an atom in the top inch of U.S. soil has 1 chance in 10 billion per year of being suspended as a dust and inhaled by a person, whereas an atom in a river, into which groundwater normally would bring the dissolved waste, has 1 chance in 10,000 of entering a human stomach.[10] Thus, an atom of waste has a much better chance of becoming absorbed into the body with food or water than with inhaled dust.

External irradiation by radioactive materials in the ground is also a lesser problem. Rock and soil are excellent shielding materials; while the waste remains buried, not a single particle of radiation from it can ever reach the surface (see Chapter 11 Appendix). Compare this with the 15,000 particles of radiation from natural sources that strike each of us every second. If radioactivity is released by groundwater, this shielding is still effective as long as it remains underground or dissolved in river water. If the radioactivity does somehow become deposited on the ground surface, it will soon be washed away or into the ground by rain. The great majority of the radiation is absorbed by building materials, clothing, or even the air. Thus, there would be relatively little exposure to the human body due to radiation originating from materials on the ground.

Quantitative calculations confirm that food and water intake is the most important mode of exposure from radioactive waste, a position agreed upon in all safety analyses.[9] Hence we will limit our consideration to that pathway. In order to estimate the hazard, we therefore start by determining the cancer risk from eating or drinking the various radioactive materials in the waste.

The method for calculating this is explained in the Chapter 11 Appendix, but the results[8] are shown in Fig. 1. For present purposes we need only look at the thick line which is above all the others in the figures. This is a plot of the number of cancers that would be expected if all of the waste produced by one plant in one year were fed to people.

If this were done shortly after burial, according to Fig. 1 we would expect 50 million cancers, whereas if it were done after 1,000 years there would be 300,000, and if after a million years there would be 2,000. Note that we did not specify how many people are involved because that does not matter.*

If nuclear power was used to the fullest practical extent in the United States, we would need about 300 power plants of the type now in use. The waste produced each year would then be enough to kill (300×50 million $=$) over 10 billion people. I have authored over 250 scientific papers over the past 35 years presenting tens of thousands of pieces of data, but that "over 10 billion" number is the one most frequently quoted. Rarely quoted, however, are the other numbers given along with it[11]: we produce enough chlorine gas each year to kill 400 trillion people, enough phosgene to kill 20 trillion, enough ammonia and hydrogen cyanide to kill 6 trillion with each, enough barium to kill 100 billion, and enough arsenic trioxide to kill 10 billion. All of these numbers are calculated, as for the radioactive waste, on the assumption that all of it gets into people. I hope these comparisons dissolve the fear that, in generating nuclear electricity, we are producing unprecedented quantities of toxic materials.

---

*For example, if feeding a quantity of waste to a thousand people would give each person a risk of $\frac{1}{10}$, we would expect ($1,000 \times \frac{1}{10} =$) 100 deaths; but if instead the quantity were fed to 10,000 people, each would eat $\frac{1}{10}$ as much and hence have a risk of only $\frac{1}{100}$, so we would expect ($10,000 \times \frac{1}{100} =$) 100 deaths, the same as before.

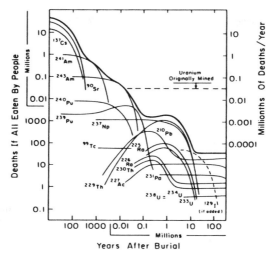

Fig. 1. Toxicity of high-level radioactive waste versus time.[8] The ordinate is the number of cancer deaths that would be expected if all the waste produced by one large nuclear power plant in one year were eaten by people. The individual curves show the toxicity of the individual radioactive species in the waste (as labeled), and the top thick curve shows their sum, the total toxicity. The horizontal dashed line shows the number of deaths expected if the uranium mined out of the ground to produce the fuel from which this waste was generated were fed to people. The scale on the right shows the deaths per year expected according to the risk analyses.

One immediate application of Fig. 1 is to estimate how much of the waste glass, converted into digestible form, would have a good chance of killing a person who eats it—this may be called a "lethal dose." We calculate it by simply dividing the quantity of waste glass, 15 tons, by the values given by the curve in Fig. 1, and the results are as follows:

shortly after burial: 0.01 oz.
after 100 years: 0.1 oz.
after 600 years: 1 oz.
after 20,000 years: 1 lb.

Lethal doses for some common chemicals are as follows[11]:

> selenium compounds: 0.01 oz.
> potassium cyanide: 0.02 oz.
> arsenic trioxide: 0.1 oz.
> copper: 0.7 oz.

Note that the nuclear waste becomes less toxic with time because radioactive materials decay, leaving a harmless residue, but the chemicals listed retain their toxicity forever.

By comparing these two lists we see that radioactive waste is not infinitely toxic, and in fact it is no more toxic than some chemicals in common use. Arsenic trioxide, for example, used as an herbicide and insecticide, is scattered on the ground in regions where food is grown, or sprayed on fruits and vegetables. It also occurs as a natural mineral in the ground, as do other poisonous chemicals for which a lethal dose is less than one ounce.

Since the waste loses 99% of its toxicity after 600 years, it is often said that our principal concern should be limited to the short term, the first few hundred years. Some people panic over the requirement of security even for hundreds of years. They point out that very few of the structures we build can be counted on to last that long, and that our political, economic, and social structures may be completely revolutionized within that time period. The fallacy in that reasoning is that it refers to our environment here on the surface of the earth, where it is certainly true that most things don't last for hundreds of years. However, if you were a rock 2,000 feet below the surface, you would find the environment to be very different. If all the rocks under the United States more than 1,000 feet deep were to have a newspaper, it couldn't come out more than once in a million years, because there would be no news to report. Rocks at that depth typically last many tens of millions of years without anything eventful occurring. They may on rare occasions be shaken around or even cracked by earthquakes or other diastrophic events, but this doesn't change their position or the chemical interaction with their surroundings.

One way to comprehend the very long-term toxicity in the waste is to compare it with the natural radioactivity in the ground. The ground is, and always has been, full of naturally radioactive materials—principally potassium, uranium, and thorium. On a long-term basis (thousands of years or more), burying our radioactive waste would increase the total radioactivity in the top 2,000 feet of U.S. rock and soil only by 1 part in 10 million. Of course, the radioactivity is more concentrated in a waste repository, but that doesn't matter. The number of cancers depends only on the total number of radioactive atoms that get into people, and there is no reason why this should be larger if the waste is concentrated than if it is spread out all over the country. In fact, quite the opposite is the case: being concentrated in a carefully selected site deep underground with the benefit of several engineered safeguards and subject to regular surveillance provides a given atom far less chance of getting into a human stomach than if it is randomly located in the ground. Incidentally, it can be shown that the natural radioactivity deep underground is doing virtually no damage to human health, so adding a tiny amount to it is essentially innocuous.

When I point out that concentration of the waste in one place does not increase the health hazard, listeners often comment that it does mean that the risks will be concentrated on the relatively small number of people who live in that area. That, of course, is true, but it is also true for just about any other environmental problem. We in Pittsburgh suffer from the air pollution generated in making steel for the whole country; citizens of Houston and a few other cities bear the brunt of the considerable health hazards from oil refineries that make our gasoline, and there are any number of similar examples. The health burden from these inequities is thousands of times larger than those from living near a nuclear waste repository will ever be.

Fig. 1 shows the toxicity of the uranium ore that was originally mined out of the ground to produce the fuel from which the waste was generated. Note that it is larger than the toxicity of the waste after 15,000 years; after that the hazard from the buried waste is less than that from the ore if it had never been mined. After 100,000 years, there is more radioactive toxicity in the ground directly above the repository (i.e., between the buried waste and the surface) than in the repository itself due to the natural radioactivity in the rock.

A schematic diagram of the waste package, as buried, is shown in Fig. 2. It is enlightening to consider the protections built into the system to mitigate the hazards. For the first few hundred years, when the toxicity is rather high, there is a great deal of protection through various time delays before the waste can escape through groundwater. First and foremost, the waste will be buried in a rock formation in which there is little or no groundwater flow, and in which geologists are as certain as possible that there will be little groundwater flow for a very long time. If the geologists are wrong and an appreciable amount of water flow does develop in the rock formation in which the waste is buried, it would first have to dissolve away large fractions of that rock—roughly half of the rock would have to be dissolved before half of the waste is dissolved. This factor might seem to offer minimal protection if the waste is buried in salt, since salt is readily dissolved in water. However, in the New Mexico area, where a repository for

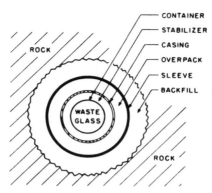

Fig. 2. Schematic diagram of buried waste package. Components are as follows: waste glass—the waste itself, converted into a glass; container—stainless steel can in which glass is originally cast; stabilizer—filler material to improve physical and chemical stability of the waste; casing—special material highly resistant to corrosion by intruding water, which it should keep out; overpack—provides additional corrosion resistance and structural stability; sleeve-liner for hole—gives structural support; backfill—material to fill space between waste package and rock, swells when wet to keep water out; if waste becomes dissolved backfill absorbs it out of the escaping water.

defense waste has been constructed, if all the water now flowing through the ground were diverted to flow through the salt, it would take a million years to dissolve the salt enclosing a repository.[11] The quantity of salt is enormous, while the water flow is not a stream but more like a dampness, slowly progressing through the ground.

The next layer of protection is the backfill material surrounding the waste package (see Fig. 2). This will be a clay, which tends to swell when wet to form a tight seal keeping water flow away from the package.

If water should penetrate the backfill, before it can reach the waste it must get past the metal casing in which the waste glass is to be sealed. Materials for this casing have been developed that give very impressive resistance to corrosion. One favorite is a titanium alloy, which was tested in a very hot (480°F versus a maximum expected repository temperature of 250°F) and abnormally corrosive solution, and even under these extreme conditions, corrosion rates were such that water penetration would be delayed for a thousand years.[12] Under more normal groundwater conditions, these casings would retain their integrity for hundreds of thousands of years. Thus, the casings alone provide a rather complete protection system even if everything else fails.

The next layer of protection is the waste form, probably glass, which is not readily dissolved. Glass artifacts from ancient Babylonia have been found in river beds, where they have been washed over by flowing river water—not just by the slowly seeping dampness which better describes groundwater conditions—for 3,000 years without dissolving away.[13] A Canadian experiment with waste glass buried in soil permeated by groundwater indicates that it will last for a hundred million years—i.e., only about $\frac{1}{100}$ millionth dissolves each year.

But suppose that somehow some of the waste did become dissolved in groundwater. Groundwater moves very slowly, typically at less than 1 foot per day—in the region of the New Mexico site it moves only about 1 inch per day and at the Nevada site which is under investigation it moves only 0.03 inches per day. Furthermore, groundwater deep underground does not ordinarily travel vertically upward toward the surface; it rather follows the rock layers, which tend to be essentially horizontal, and hence it typically must travel about 50 miles before reaching the surface. In the New Mexico area it must travel over a hundred miles and at the Nevada site at least 30 miles. Anyone can easily calculate that to travel 50 miles at 1 foot per day takes about 1,000 years, so this again gives a very substantial protection for the few hundred years that most concern us here. (Long-term effects will be considered later.)

But the radioactive material does not move with the velocity of the groundwater. It is constantly filtered out by adsorption on the rock material, and as a result it travels hundreds or thousands of times more slowly than the groundwater.[14] If the groundwater takes a thousand years to reach the surface, the radioactive materials should take hundreds of thousands or millions of years. Moreover, there are abundant opportunities along the way for waste materials to precipitate out of solution and become a permanent part of the rock.[15]

We thus have seven layers of protection preventing the waste from getting out during the first few hundred years when it is highly toxic: (1) the absence of groundwater, (2) the insolubility of the surrounding rock, (3) the sealing action of the backfill material, (4) the corrosion resistance of the metal casing, (5) the insolubility of the waste glass itself, (6) the long time required for groundwater to reach the surface, and (7) the filtering action of the rock. But even if all of these protections should fail, the increased radioactivity in the water would be easily detected by routine monitoring, and the water would not be used for drinking or irrigation of food crops; thus there would still be no damage to human health.

QUANTITATIVE RISK ASSESSMENT FOR HIGH-LEVEL WASTE

Our next task is to develop a quantitative estimate of the hazard from buried radioactive waste.[8] Our treatment is based on an analogy between buried radioactive waste and average rock. We will justify this analogy later, but it is very useful because we know a great deal about the behavior of average rock and its interaction with groundwater. Using this information in a calculation outlined in the Chapter 11 Appendix demonstrates that an atom of average rock has one chance in a trillion each year of being dissolved out of the rock, eventually being carried into a river, and ending up in a human stomach.

In the spirit of our analogy between this average rock and buried radioactive waste, we assume that this probability also applies to the latter; that is, an atom of buried radioactive waste has one chance in a trillion of reaching a human stomach each year. The remainder of our calculation is easy, because we know from Fig. 1 the consequences of this waste reaching a human stomach. All we need do is multiply the curve in Fig. 1 by one-trillionth to obtain the number of fatal cancers expected each year. This is done by simply relabeling the vertical scale, as shown on the right side of Fig. 1. For example, if all of the waste were to reach human stomachs a thousand years after burial, there would be 300,000 deaths, according to the scale on the left side of Fig. 1, but since only one-trillionth of it reaches human stomachs each year, there would be $(300,000 \div 1 \text{ trillion} =) 3 \times 10^{-7}$ deaths per year, as indicated by reading the curve with the scale on the right side of Fig. 1. This result is equivalent to one chance in 3 million of a single death each year.

What we really want to know is the total number of people who will eventually die from the waste. Since we know how many will die each year, we simply must add these up. Totalling them over millions of years gives 0.014 eventual deaths. One might ask how far into the future we should carry this addition, and therein lies a complication that requires some explanation.

From the measured rate at which rivers carry dissolved and suspended material into the ocean each year, it is straightforward to calculate that the surfaces of the continents are eroding away at an average rate of 1 meter (3.3 feet) of depth every 22,000 years. (All the while the continents are being uplifted such that on an average the surface remains at about the same elevation.) Nearly all of this erosion comes from river beds and the minor streams and runoff that feed them. However, rivers change their courses frequently; changing climates develop new rivers and eliminate old ones (e.g., 10,000 years ago the Arizona desert was a rain forest); and land areas rise and fall under geological pressures to change drainage patterns. Therefore, averaged over very long time periods, it is reasonable to assume that most land areas erode away at roughly this average rate, 1 meter of depth every 22,000 years. This suggests that the ground surface will eventually erode down to the level of the buried radioactive waste. Since the waste will be buried at a depth of about 600 meters, this may be expected to occur after about $(600 \times 22,000 =)$ 13 million years. When it happens, the remaining radioactivity will be released into rivers, and according to estimates developed in the Chapter 11 Appendix, one part in 10,000 will get into human stomachs.

Therefore, the process of adding up the number of deaths each year should be discontinued after 13 million years; that was done in calculating the number of deaths given above as 0.014. To this we must add the number of deaths caused by release of the remaining radioactivity. A simple calculation given in the Chapter 11 Appendix shows that this leads to 0.004 deaths. Adding that number to the 0.014 deaths during the first 13 million years gives 0.018 as the total number of eventual deaths from the waste produced by one power plant in one year. This is the final result of this section, 0.018 eventual fatalities. Recall now that air pollution, one of the wastes from coal burning, kills 75 people in generating the same amount of electricity; perhaps new pollution control technology may reduce this number to 15 deaths per year. Even if this comes to pass, we see that this one waste from coal burning would still be a thousand times more injurious to human health than are the high-level nuclear wastes.

Another way of understanding the results of our analysis is to consider a situation in which all of the present electricity used in the United States has been provided by nuclear power plants—about 300 would be required—continuously for millions of years. We would then expect $(300 \times 0.018 =)$ 5.4 cancer deaths per year on an average from all of the accumulated waste. Compare this with the 30,000 deaths each year we now get from coal-burning power plants.

Before closing this section, readers with a scientific bent may be interested in the justification for the basic assumption used in arriving at our estimate, the analogy between buried radioactive waste and average rock. Actually, there are three ways in which the waste may be less secure:

1. The radioactivity in the waste continues to generate heat for some time after burial, making the buried waste considerably hotter than average rock. There has been some fear that this heat may crack the

surrounding rock, thereby introducing easy pathways for ground-water to reach the waste and/or to carry off the dissolved radioactivity. This question has been extensively studied since the late 1960s, and the conclusion now is that rock cracking is not a problem unless the temperature gets up to about 650°F. The actual temperature of the waste can be controlled by the extent to which it is diluted with inert glass and by the intervals chosen for burial of the waste packages. More dilution and larger spacings result in lower temperatures. In the repositories now envisioned, these factors will be adjusted to keep the temperature below 250°F, which leaves a large margin of safety relative to the 650°F danger point. If it is decided that this margin should be larger, the easiest procedure would be to delay the burial, since the heat evolving from the waste decreases tenfold after a hundred years, and a hundredfold after 200 years. Storage facilities for waste packages have been designed, and there would be no great difficulty or expense in building them.

2. The waste glass will be a foreign material in its environment and hence will not be in chemical equilibrium with the surrounding rock and groundwater. However, chemical equilibrium is a surface phenomenon.[15] When groundwater reaches a foreign material it rapidly dissolves off a microscopically thin surface layer, replacing it with material that had been dissolved out of neighboring rocks. As this surface layer builds up in thickness, it shelters the material beneath from further interaction with the groundwater. This interaction is therefore effectively stopped after only a minute quantity of the foreign material is dissolved. The foreign material is thereafter no longer "foreign," as its surface is in chemical equilibrium with its surroundings. The fact that the waste glass is a foreign material in its rock environment is therefore not important.

3. Shafts must be dug down from the surface in order to emplace the waste packages. This might provide access routes for intrusion of water and subsequent escape of dissolved waste that are not available to ordinary rock. The seriousness of this problem depends on how securely the shafts can be sealed. That is a question for experts in sealing, and I am not one of these. However, the general opinion among the experts is that the shafts can be sealed to make them at least as secure as the original undisturbed rock.

Since we have discussed the three ways in which buried waste may be less secure than the average rock used in our analogy, it is appropriate to point out ways in which it is more secure. In the first place, burial will be in an area free of groundwater flow, whereas the rock we considered was already submerged in flowing groundwater. Second, it will be in a carefully chosen rock environment, whereas the rock in our analogy is in an average environment; it might, for instance, be in an area of extensive rock fracturing. Third, the waste package will be in a special casing which provides extreme resistance to water intrusion, a protection not available to average rock; this casing should provide adequate protection even if everything else imaginable goes wrong. And finally, if any radioactive waste should escape and get into food and water supplies, it would very probably be detected by routine monitoring programs in plenty of time to avert any appreciable health consequences.

In balancing the ways in which buried waste is more secure and less secure than the average rock submerged in groundwater with which we compared it, it seems reasonable to conclude that our analogy is a fair one. If anything, it overestimates the dangers from the waste.

It should be pointed out that there are other methods than the one presented here for quantifying the probability for an atom of material in the ground to enter a human stomach. For example,[11] we know the amounts of uranium and radium in the human body from measurements on cadavers, and we know the amounts in the ground. From the ratios of these, we can estimate the probability per year for an atom of these elements in the ground to enter a human body. This gives a result in good agreement with that of the calculation we presented.

Sometimes when I complete a demonstration like the one above of the safety of buried radioactive waste, a listener complains that the demonstration is invalid because we don't know how to bury waste. How anyone can possibly be so naive as to believe that we don't know how to bury a simple package is truly a challenge to my imagination. Many animals bury things, and people have been burying things for millions of years. Any normal person could easily suggest several different ways of burying a waste package. Why, then, are we devoting so much time and effort to developing a waste burial technology? The problem here is to decide on the best way of burying the waste to maximize its security. Whether it is worthwhile to go to such pains to optimize the solution is highly questionable, especially since we have shown that even the haphazard burial of average rock by natural processes would provide adequate safety. But the public's irrational worries about the dangers have forced our government to spend large sums of money on the problem. Actually, we are paying for it in our electricity bills. Fortunately, it only increases our bills by one percent.

LONG-TERM WASTE PROBLEMS FROM CHEMICAL CARCINOGENS

In order to appreciate the meaning of the result obtained in the last section — 0.018 eventual cancer deaths from each plant-year of operation — it is useful to put it in perspective by comparing it with the very long-term cancer risks from the buried wastes produced by other methods of generating electricity. Here we consider certain types of chemicals that cause cancer[8]; other examples will be given in the next chapter.

The only reason buried radioactive waste has a calculable health effect is because we are using a linear dose-response relationship, assuming, for example, that if 1 million mrem causes a given risk, 1 mrem will cause a millionth of that risk. This linear relationship does not apply to most toxic chemicals, like carbon monoxide poisoning, for which there is a "threshold" exposure required before there can be any harm. However, it is now widely believed that a linear, no-threshold relationship is valid for chemical carcinogens (i.e., chemicals that can induce cancer) just as for radiation. Basically, the reason for this thinking in both cases is that cancer starts from injury to a single molecule on a single chromosome in the nucleus of a single cell, often induced by a single particle of radiation or by a single carcinogenic molecule. The probability for this simple process is not dependent on whether or not other carcinogenic molecules are causing injury elsewhere. Thus the probability of cancer is simply proportional to the number of carcinogenic molecules taken into the body, as described by a linear relationship. This linear, no-threshold, dose-response relationship for chemical carcinogens has now been officially accepted by all U.S. Government agencies charged with responsibilities in protecting public health, such as the Environmental Protection Agency, the Occupational Safety and Health Administration, and the Food and Drug Administration.

The particular carcinogens of interest in our discussion are the chemical elements cadmium, arsenic, beryllium, nickel, and chromium. There is a considerable body of information indicating that all of these cause cancer and enough quantitative information to estimate the risk per gram of intake. All of them are present in small amounts in coal. Therefore, when coal is burned, they are released and eventually, by one route or another, end up in the ground. Small amounts of each of them are also contained in food. All such materials in food, of course, are taken in from the ground through the roots of plants.

As in the case of radioactive waste, the principal problem in evaluating the hazard is to estimate the probability for an atom of these elements in the ground eventually to reach a human stomach. However, the problem is simpler here because the toxicity in these carcinogenic elements does not decay with time, as does that of radioactive waste. Since some quantity of these elements is ingested by humans every year, if an atom remains in the ground indefinitely, eventually after a long enough time, it has a good chance of reaching a human stomach. However, there are other processes for removing materials from the ground. The most important of these is erosion of soil, with the materials being carried by rivers into the oceans. The probability for an atom in the ground to enter a human stomach is therefore the probability for this to happen before it is washed into the oceans. This is just the ratio of the rate for entering human stomachs to the rate of removal by erosion.

The rates for both of these are known. From chemical analyses of food and the quantity consumed each year, we can calculate the rate at which cadmium, for example, enters human stomachs in the United States. From the rate at which American soil is washed into oceans and the measured abundance of cadmium in soil, we can calculate the rate at which cadmium is carried into oceans. We thus can determine the probability for an atom of cadmium in the soil to enter a human stomach. It turns out to be surprisingly high, about 1.3%. It is 5 times smaller for arsenic, 20 times lower for beryllium and nickel, and 60 times lower for chromium (cadmium atoms are more easily picked up by plant roots because of their chemical properties).

Since we know the quantities of each of these carcinogenic elements deposited into the top layers of the ground as a result of coal burning, the probability for them to be transferred from the ground into humans, and their cancer risk once ingested, it is straightforward to estimate the number of deaths caused by coal burning. The result is about 50 deaths due to one year's releases from one large plant. These deaths are spread over the time period during which the top few meters of ground will be eroded away, about 100,000 years.*

If one takes a much longer-term viewpoint, these deaths would have eventually occurred even if the coal had not been mined and burned, because after several millions of years, erosion would have brought this coal near the surface, where its cadmium, arsenic, beryllium, chromium, and nickel could have been picked up by plant roots and gotten into food. However, the fact that the coal was mined means that its turn near the surface is taken by other rock which, we assume, has the average amount of these carcinogenic elements in it. This carcinogenic material represents a net additional health

---

*We have shown that the average erosion rate is 1 meter every 22,000 years. We imply here that most of the material ends up within something like 4.5 meters (15 feet) of the surface $(4.5 \times 22,000 = 100,000)$.

risk to humankind due to the use of coal. Taking this into account, the final result is that there would be a net excess of 70 deaths. Note that this is thousands of times larger than the 0.018 deaths from high-level radioactive waste.

The chemical carcinogens also have an impact on the long-term health consequences of electricity produced by solar energy. We have already pointed out that producing the materials for deployment of a solar array requires about 3% as much coal burning as producing the same amount of electricity by direct coal burning. The quantity of chemical carcinogens released in the former process is therefore 3% of those released in the latter, so in the very long term we may expect about 2 eventual deaths (3% of 70) from an amount of solar electricity equivalent to that produced by a large nuclear or coal-fired power plant in one year. This is a small number, but it is still a hundred times the 0.018 deaths from nuclear plant waste.

But there may be much more important problems. A prime-candidate material for future solar cells is cadmium sulfide. When these cells deteriorate, that material will probably be disposed of in one way or another into the ground. The cadmium introduced into the ground in this way is estimated to eventually cause 80 deaths for the quantity of electricity produced by a large nuclear plant in one year. Most of the cadmium used in this country is imported. Thus, in contrast to the situation for coal, it would not have eventually reached the surface of U.S. soil without this technology. Again we encounter consequences thousands of times higher than the 0.018 deaths from radioactive waste.

SHOULD WE ADD UP EFFECTS OVER MILLIONS OF YEARS?

Many may be bothered by the idea of adding up effects over millions of years as we have been doing in this chapter. Personally, I agree with them. The idea of considering effects over such long time periods was introduced by opponents of nuclear power, who often insist that it be done. The analyses presented above show that such a procedure turns out very favorably for nuclear power. However, there are many good reasons for confining our attention to the "foreseeable future," which is generally interpreted to be 100-1,000 years. Several groups, including most government agencies, have adopted 500 years as a reasonable time period for consideration, so we will use it for this discussion.

The reason for ignoring lives that may be lost in the far future is not because those lives are less valuable than our own or those of our closer progeny. There surely can be no moral basis for any such claim. But there are at least three reasonable and moral bases for ignoring the distant future.[16]

One such basis is the improvement in cancer-cure rates. In recent years, they have been improving by about 0.5% per year. It is unlikely that cure rates will continue to improve at this rate. More probably they will stagnate before long, or on the other hand there may be a breakthrough leading to a dramatic improvement. But most knowledgeable people agree that there is a good chance that cancer will be a largely curable disease in 500 years. The deaths we calculate to occur after that time will therefore probably never materialize.

The second basis for ignoring deaths calculated to occur in the distant future is what I call the "trust fund" approach. We showed in Chapter 8 that there are many ways that money can be used to save lives at a rate of at least one life saved per $100,000 spent. But even suppose that so many improvements are instigated that it will cost $1 million to save a life in the distant future (all our discussions discount inflation). There is a continuous record extending back 5,000 years of money always being able to draw at least 3% real (i.e., discounting inflation) interest. Each dollar invested now at 3% interest is worth $2.5 million after 500 years and will therefore be capable of saving more than one life. It is therefore much more effective in saving far future lives to set up a trust fund to be spent in their time for that purpose rather than for us to spend money now to protect them from our waste. One might wonder about the mechanics and practicality of setting up a trust fund to be transmitted to future generations, but this is no problem. By overspending, we are now building up a public debt that they will have to pay interest on, so rather than actually set up a trust fund, we need only reduce our spending and thereby leave future generations with more money to use for lifesaving. By not spending profligately on waste management, we are doing just that.

A third basis for ignoring the distant future in our considerations is the "biomedical research" approach. If we don't feel comfortable with the "trust fund" approach of putting aside money to be spent in the future for saving lives, we can save future lives by spending money today on biomedical research. One study concludes that the $68 billion spent on such research between 1930 and 1975 is now saving 100,000 lives per year in the United States, one life per year for every $680,000 spent. Another study estimates that the $20 billion spent in 1955-1965 is now saving 100,000 lives per year, or one life per year for every $200,000 spent. Age-specific mortality rates in the United States have been declining steadily in recent years; if one-fourth of this decline is credited to biomedical research, the latter is now saving one life per year for every $1 million spent. As the easier medical problems are solved and the more difficult ones are attacked, the cost goes up. Let us therefore say that for the near future, it will cost $5 million to save one life per year in the United States, not to mention that it will save lives in other countries. Over the next 500 years, saving one life per year will avert a total of 500 deaths, a cost of $10,000 per life saved. If we extend our considerations beyond 500 years, the price drops proportionally. Since we are now spending many millions of dollars per future life saved in handling our radioactive waste, it would clearly be much more beneficial to future generations if we put this money into biomedical research.

The straightforward way to implement this alternative would be for the nuclear industry to contribute money to biomedical research rather than to improved long-term security of nuclear waste, but even that may not be necessary. The nuclear industry pays money to the government in taxes, and the government spends roughly an equal sum of money to support biomedical research. All that is necessary to implement an equitable program from the viewpoint of future citizens is that 0.1% of this cash flow from the nuclear industry to government to biomedical research be redefined as support of biomedical research by the nuclear industry. If it would make people happier, the government tax on the nuclear industry could actually be raised by 0.1% and this money could be directly added to the government's biomedical research budget. No one would notice the difference since it is much less than a typical annual change in these taxes and budgets.

As a result of any one of the above three rationales, or a combination of them, many people, including myself, feel that we should worry only about effects over the next 500 years. For their benefit we now develop estimates of the health effects expected during that time period, from one year's operation of a large power plant[8]:

For the high-level radioactive waste, we take the number of fatalities each year from using the scale on the right side of Fig. 1 and add them up over the first 500 years. The result is 0.003 deaths, most of them in the first 100 years. However, this ignores all the time delays and special protections during this early time period that were outlined above: geologists feel confident that no groundwater will intrude into the waste repository for at least a thousand years; the corrosion-resistant casing is virtually impregnable for several hundred years; it takes groundwater a thousand years to get to the surface; and so on. Moreover, since it is quite certain that some sort of surveillance would be maintained during this period, any escaping radioactivity would be detected in plenty of time to avert health problems. These factors reduce the dangers at least 30-fold, to no more than 0.0001 deaths. This is close to a million times less than the number of deaths due to air pollution from coal burning. The chemical carcinogens released into the ground in coal burning do about 1% of their damage during the first 500 years; we may consequently expect 0.5 deaths in that period. With a 500-year perspective, there is no need to consider the much later effects of the coal reaching the surface by erosion if it had not been mined. We see that the effects of chemical carcinogens from coal burning are 5,000 times worse than those of nuclear waste $(0.5 \div 0.0001)$.

By similar reasoning, the chemical carcinogens from solar electricity will cause 0.015 deaths due to the coal used in making materials for it; if cadmium sulfide solar cells are used, they will cause 0.8 deaths. Again these are far greater than the 0.0001 deaths due to the waste from nuclear power.

SITING A WASTE REPOSITORY[17,18]

As a consequence of public fear, siting a repository for high-level waste has become a very difficult and expensive process. The NIMBY syndrome — Not in my back yard — receives its fullest expression here because there is little obvious benefit and enormous perceived harm to the area that accepts the waste. There is even a stigma attached to being the place where the nation's "atomic garbage" is sent. The issue is made to order for local environmental groups. By opposing the repository they get lots of media coverage, which is the essential ingredient needed for them to prosper. The issue is also made to order for local politicians who can garner lots of votes by posing as the defender of the home territory against the intrusion of the federal government and the commercial interests of the rest of the nation. There are few votes to be gained by supporting the local siting of a repository.

The U.S. Department of Energy (DOE), which is charged with the responsibility for managing the waste, has faced this difficult situation by adopting the position that repository site selection is strictly a scientific problem, requiring a tremendous research effort which it is more than happy to undertake. The money for it is available from the 0.1 cent per kilowatt-hour tax on nuclear electricity. Through 1989, $6 billion had already been collected. Managing a large research program is clearly advantageous to the careers of those in charge, and it also delays the pain of making a decision. The longevity of individuals in top government positions is such that they personally will probably never have to make a final decision—for a government bureaucrat that is the ideal situation.

As a result of these factors, selection of a repository site has been painfully slow. The current status is that a tentative site has been selected, under Yucca mountain in Nevada, 100 miles northwest of Las Vegas, and research is proceeding to determine whether it is suitable. The research *plan,* published in December 1988, runs more than 6,300 pages and cost $30 million! The research project itself was estimated to take 7 years and cost $2 billion. It was to include drilling and blasting two shafts 1,000 feet deep to the repository level, and carving out an underground laboratory there for carrying out experiments that simulate the underground environment. In addition, there were to be over 300 drill holes to explore the geology and hydrology beyond the laboratory.

A year later, the new Secretary of Energy decided that the entire research plan should be revised. As part of the new procedure, the repository will not become operational until 2010.

Yucca mountain is in a very barren desert region, adjacent to the nuclear weapons testing facility, an area larger than the state of Rhode Island known as the Nevada Test Site (NTS). The NTS itself was originally part of, and along with Yucca mountain is still surrounded by, a much larger U.S. Air Force bombing and gunnery range.

Because of the NTS, radioactivity is not a stranger to the area. About 100 nuclear bomb tests were conducted in the atmosphere there between 1951 and 1963, and since 1963, 600 more announced tests have been conducted underground. Additional tests were not announced to the public. In the late 1980s, an average of 14 tests per year were announced. Each underground bomb test can be considered as a miniature high-level waste repository.

One can stand on high ground overlooking Frenchman's Flat and see many dozens of craters, each caused by surface subsidence above a cavity caused by an underground nuclear explosion. One can see heavily damaged buildings and bridges exposed to the earlier tests in the atmosphere. The most spectacular sight is Sedan crater, 1,280 feet in diameter and 320 feet deep, created in a test of the potential use of nuclear bombs for large excavation projects—that one bomb removed 12 million tons of earth. But most testing is now conducted in "rooms" leading off of tunnels drilled into the sides of mountains.

Yucca mountain is several miles away from the part of NTS used for bomb testing. The NTS areas nearby have been used for testing nuclear reactors developed for spaceship propulsion, a project that has now been abandoned. Any possible effects of the bomb testing on the security of the waste repository will be carefully considered.

The rock under Yucca mountain is welded tuff, a volcanic lava which oozed out of the earth 13 to 18 million years ago and remained at high temperature and out of contact with air for a long time. This caused it to become tightly welded into a dense, nonporous rock.

Yucca mountain is one of the driest places in the United States, with only 6 inches of rainfall per year, only 5% of which (0.3 inches per year) soaks into the ground. The water table is more than 2,500 feet deep, so that the 1,200-foot-deep repository will be 1,300 feet above it. Only a tiny amount of moisture can therefore contact the buried waste. The rock below the repository is especially rich in zeolites, materials with chemical properties that make them efficient at absorbing dissolved radioactive materials out of water passing through. It is estimated that water percolating down would take 2,000-8,000 years to travel this 1,200 feet from the repository to the water table.

If any radioactive waste should be dissolved out, and if it should somehow escape adsorption in the zeolites and reach the water table, consequences would still be minimal. Groundwater in the region moves especially slowly, only about 1 foot per year! The nearest place it could reach the surface is 30 miles away—150,000 years at 1 foot per year—in the Amargosa desert, which is a closed drainage basin—water flowing into it stays there until it evaporates. Note that the 1 foot per year flow rate applies to the water; as explained previously, materials dissolved in it move hundreds or thousands of times more slowly. For them to travel 30 miles would take many millions of years!

In summary, there is virtually no moisture to dissolve the waste, which is in any case in insoluble form and protected by casings and backfill material. If any material is dissolved, the only escape path is downward 1,300 feet to the top of the water table through rock that is highly efficient at adsorbing out the dissolved materials. If it should reach the water table it would take many millions of years to reach the surface, and that would leave it trapped in the middle of a virtually uninhabited desert with no way to get out. Note that this is very much safer than the situation assumed in our risk analysis (in Chapter 11 Appendix) for average underground conditions in which the waste is submerged in groundwater (i.e., below the top of the water table) that moves at a velocity of 1 foot per *day,* and flows into a river through a region of U.S. average population.

There are other matters to consider, including the potential effects of climate change, volcanoes, and earthquakes. All of these now seem to be of minimal concern. The closest major earthquake was 90 miles to the west in 1872, and there has been no major faulting or folding of rocks for over 10 million years. The climate has been arid for over 2 million years, and there have been no volcanoes for several million years. But investigating these matters is an important part of the research program.

In spite of all this, the Las Vegas area, which has lived with nuclear bomb testing at similar or closer distances for nearly 40 years, is alive with opposition, including citizens' alliances, residents' coalitions, governor's commissions, study committees, and the like. They are composed of ordinary people—secretaries, teachers, iron workers, printers, lawyers, real estate brokers. In accordance with the law, some of these are funded by the DOE out of funds from the tax on nuclear electricity. They say that, in accepting the nuclear bomb testing facility, Nevada has done its share, so some other state should accept the nuclear waste.

Nevada's governors and senators have opposed the repository. The state legislature passed resolutions opposing it, refusing to yield jurisdiction and withholding consent. It even passed a law making it a crime for anyone to store high-level radioactive waste in the state punishable by fines and imprisonment. That law is being tested in the courts. The state has filed several law suits against the DOE, and the legal situation is cluttered with counter-suits and appeals.

Some Nevadans favor the repository as a diversification of industry for the state, which is highly dependent on tourism and gaming. Some fears have been expressed that other industries may be deterred from moving to the area, but in 1988-89, 50 businesses moved into Nevada and none raised the repository issue as a concern. Some Nevadans are more concerned with negotiating the best possible deal in terms of payments from the DOE for accepting the repository. There is every indication that the state will be well compensated.

If the legal problems can be worked out, and if the technical evaluation of the site does not encounter serious problems, the Yucca mountain repository will be built and operating by about the year 2010. The cost will be tremendous, but it will be covered by the 0.1 cent per kilowatt-hour tax which adds only 1% to the cost of nuclear electricity.

WHY THE PUBLIC FEAR?

If the picture I have presented in this chapter is correct, does not omit important aspects of the problem, and is not misleading, it is difficult for me to understand why people worry about the dangers of high-level radioactive waste. The material presented here has been published in the appropriate scientific journals for some time. It had been approved by scientific referees and editors to get there. There have been no critiques of it offered. I have presented it at international, national, and regional scientific conferences, and in seminar talks at several major universities, all without receiving substantive criticisms or adverse comments. I therefore know of no evidence that my treatment is not a correct and valid assessment of the problem based on scientific analyses.

There have been other scientific analyses of the high-level radioactive waste hazards using very different approaches that I believe to be less valid than mine, but they come out with rather similar results.[9] They also find the health effects to be trivial.

Why, then, is there so much opposition? What is the basis for the public worry about this problem? There have been a few scientific papers that concoct very special scenarios in which substantial—but still far from catastrophic—harm to public health might result, but they never attempt to estimate the likelihood of their scenarios. All of them are obviously extremely improbable. Note that in my treatment the probabilities of all scenarios are automatically taken into account as long as they apply equally to average rock.

Nevertheless, there can be no question but that the fear abounds. A segment on a network TV morning show opened with the statement "[this waste] could possibly contaminate the environment. The result, of course, would be too horrible to contemplate; it could eventually mean the end of the world as we now know it." I cannot imagine any situation in which this last statement is meaningful short of feeding all of the people in the world directly with the waste. As has been pointed out, there are many other substances that would serve with greater effectiveness if this were done.

Probably the best evidence for this fear is that our government is willing to spend huge sums of money on the problem. The fear is surely there. Perhaps the most important reason for it is that disposal of high-level waste is often referred to as an "unsolved problem." When people say this to me, I ask whether disposal of the waste from coal burning—releasing it as air pollution, killing about 30,000 Americans each year—is a "solved problem." They usually say "no" but add that air pollution is being reduced. Actually we can hardly hope to reduce it to the point of killing fewer than 5,000 Americans per year. It is therefore difficult to understand how that converts it into a "solved problem." When I point out that the radioactive waste problem is not unsolved, since there are many known solutions, the usual retort is "then why aren't we burying it now?" The answer to that question is we are in the process of choosing the best solution. This is about as far as the discussion goes, but the real difficulty is that it rarely goes that far. The great majority of people are still hung up on "it's an unsolved problem."

I like to point out that nobody is being injured by the high-level waste, and there is no reason to believe that anyone will be injured by it in the foreseeable future—compare this with the 30,000 deaths per year from the wastes released in coal burning. The usual reply is "how do you know that nobody is being injured?" The answer is that we constantly measure radiation doses in and around nuclear facilities and throughout our environment, and nobody is receiving any such doses from high-level waste. If there is no radiation dose, there can be no harm. I don't believe any scientist would argue this point, but I'm sure most people continue to believe that high-level waste (or the spent fuel from which it is to be derived) actually is doing harm now.

The real difficulty with public understanding of the high-level waste problem is that the scientists' viewpoint is not being transmitted to the public. Transmitting information from the scientific community to the public is in the hands of journalists, who have chosen not to transmit on this question. I'd hate to speculate on their motives, but they are doing great damage to our nation. These are the same journalists who constantly trumpet the claim that the government is suppressing information on nuclear energy. What more despicable suppression of information can there be than refusing to transmit the truth about important questions to the public? It's every bit as effective as lying.

Chapter 12 / MORE ON
RADIOACTIVE WASTE

While high-level waste has received the lion's share of the public's attention, it is not the only waste problem linked with nuclear power. In fact, analyses indicate that it is not even the most important one. Several other nuclear waste issues must be addressed and we will consider them here.

In terms of numbers of deaths, the most important waste problems arise from the radioactive gas radon. We begin with its story.

RADON PROBLEMS

Uranium is a naturally occurring element, present in small quantities in all rock and soil, as well as in materials derived from them such as brick, plaster, and cement. It is best known as a fuel for nuclear reactors, but quite aside from the properties that make it useful for that purpose, it is naturally radioactive. This means that it decays into other elements, shooting out high-speed particles of radiation in the process. The residual atoms left following its decay are also radioactive, as is the residue from decay of the latter, and so on, until the chain is terminated after 14 successive radioactive decays. One step in this chain of decays is radon, which is of very special interest since it is a gas. An atom of radon, behaving as a gas, has a tendency to move away from the location where it was formed and percolate up out of the ground into the air. Since uranium is in the ground everywhere, atoms of radon and the radioactive elements into which it decays, known as "radon daughters," float around in the air everywhere and are thus constantly inhaled by people. Since it is a gas, radon itself is very rapidly exhaled. But the elements into which it decays are not gases; hence, they tend to stick to the surfaces of our respiratory passages, exposing the latter to radiation, and thereby inducing lung cancer.

Scientists know a great deal about this problem because of several situations in which miners were exposed to very high levels of radon in poorly ventilated mines.[1] In uranium mines, especially, there is often an unusually high concentration of uranium in the surrounding rock. The radon evolving from it percolates out into the mine where it remains for some time, since its escape routes are limited. The best study of this problem involved a group of 4,000 uranium miners who worked in the Colorado plateau region between 1945 and 1969. Among them, 256 men have died from lung cancer, whereas only 59 lung cancers would have been expected from an average group of American men. When this situation was recognized in the late 1960s, drastic improvements were introduced in the ventilation of these mines, reducing radon levels about 20-fold, and making present radon exposure one of the less important of the many significant risks associated with mining.

From the studies of lung cancer among miners, we can estimate the risks in various other radon exposure situations. The most important of these is in the normal environmental exposure we all receive, especially in our homes, which are often poorly ventilated. Radon percolates up from the ground, usually entering through cracks in the floor or along pipe entries. In some areas it enters homes with water drawn from underground wells. As a result of its being trapped inside homes, radon levels indoors are normally several times higher than outdoor levels.

Using the data on miners and applying the linear hypothesis leads to an estimate that this environmental radon exposure is now causing about 14,000 fatal lung cancers each year in the United States,[1] several times more fatalities than are caused by all other natural radiation sources combined.

Some people, trying to conserve fuel, carefully seal windows and doors to reduce air leakage from houses. This traps radon inside for longer time periods and therefore increases the radon level in the house. If everyone sealed their houses in accordance with government recommendations, many thousands of additional fatalities each year would result. Recalling from Chapter 8 that most scientists estimate the radiation consequences of a full nuclear power program as less than 10 fatalities per year (even the opponents of nuclear power estimate only a few hundred fatalities per year), we see that conservation is a far more dangerous energy strategy than nuclear power from the standpoint of radiation exposure!

These environmental radon problems have no direct connection with nuclear power, but there are connections in other contexts.[2] The most widely publicized of these is the problem of uranium mill tailings. When uranium ore is mined, it is taken to a nearby mill and put through chemical processes to remove and purify the uranium. The residues, called the tailings, are dissolved or suspended in water that is pumped into ponds. When these ponds eventually dry out, they leave what ostensibly are piles of sand. However, this sand contains the radioactive products from the decay of uranium; only the uranium itself has been removed. The most important of these decay products is thorium-230, which has an average lifetime of 110,000 years and decays into radium, which later decays into radon. In fact, most of the radon generated over the next 100,000 years will come from uranium that has already decayed into thorium-230, so removing the uranium has little effect on radon emissions over this 100,000 year period. But a very important effect of mining the uranium is that the source of this radon was removed from underground where the ore was originally situated, to the surface where the mill tailings are located. This allows far more of it to percolate into the air to cause health problems.

A quantitative calculation of this effect indicates that the mill tailings produced in providing a 1-year fuel supply for one large nuclear power plant will cause 0.003 lung cancer deaths per year with its radon emissions. This may not seem large, but it will continue for about 100,000 years, bringing the eventual fatality toll to something like 220! Fortunately, there are measures that can be, and are being, taken to alleviate this problem. If the tailing piles are covered with several feet of soil, nearly all of the radon will decay away during the time it takes to percolate up through the cover—the average lifetime of a radon atom is only 5.5 days. The law now requires that radon emissions from the tailings be reduced 20-fold by covering them with soil (or with other materials). This lowers the eventual fatality toll from 220 to 11. Even this is hundreds of times higher than the 0.018 fatality toll from the high-level waste. Anyone who worries about the effects of radioactive waste from the nuclear industry should therefore worry much more about the uranium mill tailings than about the high-level waste. However, for some reason the latter has received the great majority of the publicity and hence the most public concern. As a result, the government research program on controlling mill tailings is only a small fraction of the size of the program on high-level waste. The dollars are spent in response to public concern, rather than where the real danger lies.

By far the most important aspect of radon from uranium mining is yet to be discussed.[2] When uranium is mined out of the ground to make nuclear fuel, it is no longer there as a source of radon emission. This is a point which has not been recognized until recently because the radon that percolates out of the ground originates largely within 1 meter of the surface; anything coming from much farther down will decay away before reaching the surface. Since the great majority of uranium mined comes from depths well below 1 meter, the radon emanating from it was always viewed as harmless. The fallacy of this reasoning is that it ignores erosion. As the ground erodes away at a rate of 1 meter every 22,000 years, any uranium in it will eventually approach the surface, spending its 22,000 years in the top meter, where it will presumably do great damage. The magnitude of this damage is calculated in the Chapter 12 Appendix, where it is shown that mining uranium to fuel one large nuclear power plant for one year will eventually save 420 lives! This completely overshadows all other health impacts of the nuclear industry, making it one of the greatest lifesaving enterprises of all time if one adopts a very long-term viewpoint.

Before we can count these lives as permanently saved, we must specify what is to be done with the uranium, for only a tiny fraction of it is burned in today's nuclear reactors. There are two dispositions that would be completely satisfactory from the lifesaving viewpoint. The preferable one would be to burn it in breeder reactors and thereby derive energy from it. An easy alternative, however, is to dump it into the ocean. Uranium remains in the ocean only for about a million years before settling permanently into the bottom sediments. All uranium in the ground is destined eventually to be carried by rivers into the ocean and spend its million years therein. From a long-range viewpoint it makes little difference to the health of humans or of sea animals if it spends that time in the ocean now or a million years in the future. But by preventing it from having its 22,000 year interlude within 1 meter of the ground surface, we are saving numerous human lives from being lost due to radon.

If one adopts the position that only effects over the next 500 years are relevant, there is still an important effect from uranium mining because about half of all uranium is surface mined. Approximately 1% of this comes from within 1 meter of the ground surface where it is now serving as a source of radon exposure. It is shown in the Chapter 12 Appendix that eliminating this source of exposure by mining will save 0.07 lives over the next 500 years, but in the meantime, radon escaping from the covered tailings will cause 0.07 deaths. Thus, the net health effect of the radon from uranium mining is essentially zero over the next 500 years.

We still have one more radon problem to discuss, namely, the radon released in coal burning. Coal contains small quantities of uranium; when the coal is burned, by one route or another, this uranium ends up somewhere in the ground. Again, the problem is complicated by the fact that, if the coal had not been mined, erosion would eventually have brought the coal with its uranium to the surface anyhow. The final result, derived in the Chapter 12 Appendix, is that the extra radon emissions caused by the burning of coal in one large power plant for one year will eventually cause 30 fatal lung cancers. This toll, like so many others we have encountered here, is thousands of times larger than the 0.018 deaths caused by the high-level waste produced in generating the same amount of electricity from nuclear fuel.

As discussed previously, solar electricity burns 3% as much coal as would be needed to produce the same amount of electricity by direct coal burning. Solar electricity should therefore be charged with (3% of 30 =) one death per year from radon. This again, is far larger than the 0.018 deaths per year from high-level radioactive waste.

ROUTINE EMISSIONS OF RADIOACTIVITY

Nuclear power plants, and more importantly, fuel-reprocessing plants, routinely emit small quantities of radioactive material into the air and also into nearby rivers, lakes, or oceans. Opponents of nuclear power made a rather big issue of this in the early 1970s, but there has been relatively little publicity about it in recent years. The quantities of radioactivity released are limited by several Nuclear Regulatory Commission requirements, including requiring that equipment be installed to reduce these emissions if doing so will result in saving one life from radiation exposure for every $4 million spent.[3] Since new and improved control technologies are steadily being developed, these emissions have been reduced considerably since the early 1970s, and that trend is continuing. There are also NRC requirements on maximum exposure to any individual member of the public, limiting his or her exposure to 5 mrem per year, less than 2% of that from natural radiation; 0.5% is typical for those living very close to a nuclear plant. This gives them a risk equivalent to that of driving 3 extra miles per year, or of crossing a street one extra time every 4 months.

The principal emissions of importance from the health standpoint are radioactive isotopes of krypton and xenon (Kr-Xe), gases which are very difficult to remove from the air; tritium (T), a radioactive form of hydrogen that becomes an inseparable part of water, used in such large quantities that it would be impractical to store it until the tritium decays away; and carbon-14 (C-14), which becomes an inseparable part of the omnipresent gas carbon dioxide. Of these, only C-14 lasts long enough to irradiate future generations.

According to a recent study by the United Nations Scientific Committee on Effects of Atomic Radiation,[4] the releases associated with 1 year of operation of one power plant can be expected to cause 0.06 deaths from Kr-Xe, 0.016 from T, and 0.23 from C-14 over the next 500 years and 1.6 deaths over tens of thousands of years. This gives a total of 0.25 over the next 500 years and 1.6 deaths eventually. Nearly all of these effects are a consequence of fuel reprocessing, and they are spread uniformly over the world's population. Thus only a few percent of the deaths would be in the United States. There are improved technologies for reducing these emissions drastically, and regulations for implementing some of them have already been promulgated, but there is no rush to complete them since we are not doing reprocessing. If and when we do, it is reasonable to expect emissions to be reduced about 5-fold. In our summary we will assume that to be the case.

Note again that these effects are many times larger than the 0.0001 deaths during the first 500 years and the 0.018 eventual deaths from high-level waste. They are also larger than the 0.02 deaths estimated by the Reactor Safety Study as the average toll from reactor accidents. Here again, we see that public concern is driven by media attention and bears no relationship to actual dangers.

LOW-LEVEL RADIOACTIVE WASTE

In order to minimize the releases of radioactivity into the environment as discussed in the last section, there is a great deal of equipment in nuclear plants for removing radioactive material from air and water by trapping it in various types of filters. These filters, including the material they have collected, are the principal component of what is called low-level radioactive waste, the disposal of which we will be discussing here. Other components are things contaminated by contact with radioactive material, like rags, mops, gloves, lab equipment, instruments, pipes, valves, and various items that were made radioactive by being in or very near the reactor, where they were exposed to neutrons. Not all of the low-level radioactive waste is from the nuclear industry. Some is from hospitals, research laboratories, industrial users of radioactive materials, and the like. These make up about 25% of the total.

In general, the concentration of radioactivity in low-level waste is a million times lower than in high-level waste—that is the reason for the name—but the quantities in cubic feet of the former are thousands of times larger. It is therefore neither necessary nor practical to provide the low-level waste with the same security as the high level. The low-level waste is buried in shallow trenches about 20 feet deep in commercially operated burial grounds licensed by the federal or state governments.

Since its potential for doing harm is relatively slight, until recently this low-level waste was handled somewhat haphazardly. There was little standardization in packaging, in handling, or in stacking packages in trenches, and little care in covering them with dirt. As a result, trenches sometimes filled with rainwater percolating down through the soil, which then dissolved small amounts of the radioactivity. The caretakers regularly pumped this water out of the trenches and filtered the radioactivity out of it, allowing no radioactivity to escape.

The situation was radically changed by two very innocuous but highly publicized incidents during the 1970s. In an eastern Kentucky burial ground, a place called Maxey Flats, tiny amounts of radioactivity were found off site. The quantities were so small that no one could have received as much as 0.1 mrem of radiation (1 chance in 40 million of getting a cancer). The publicity was enormous at the time. The January 18, 1976, issue of the Washington Star carried a story headlined "Nuclear Waste Won't Stay Buried," which began with "Radioactive wastes are contaminating the nation's air, land, and water." When the head of the U.S. Energy Research and Development Agency (predecessor of Department of Energy) testified for his Agency's annual budget, the first 25 minutes of questions were about Maxey Flats, and a similar pattern was followed with the chairman of the Nuclear Regulatory Commission.

After another congressional committee was briefed about the problem, its chairman stated publicly that it was "the problem of the century." I pointed out to him that due to the high uranium content of the granite in the congressional office building, his staff was being exposed to more excess radiation every day than anyone had received in toto from the Kentucky incident. He made no further alarmist statements, but as a result of all the attendant publicity the burial ground was closed.

The other incident occurred in a western New York State burial ground where there was a requirement that permission be granted from a state agency to pump water out of the trenches. In one instance, this permission was somehow delayed, in spite of urgent warnings from the site operators. As a result, some slightly contaminated water overflowed, with completely negligible health consequences—the largest doses were 0.0003 mrem—leading to widespread adverse publicity and closing of the burial ground.

A television "documentary" on the Kentucky burial ground caused a lot of problems. It showed a local woman saying that the color of the water in the creek had changed, but it did not point out that any scientist would agree that this could not possibly be due to the radioactivity. Moreover, the producer was told that the coloring was caused by bulldozing operations nearby. The same program showed a local farmer complaining that his cattle were sick— "Hair been turning gray, grittin' their teeth, and they're a-dying, going up and down in milk." There was no mention of the facts that a veterinarian had later diagnosed the problem as a copper and phosphorus deficiency, the cattle had been treated for this deficiency and had recovered, and that the TV crew had been informed about this long before the program was aired.

A large portion of a newspaper feature story was devoted to the story of a woman telling about how she was dying of cancer due to radiation caused by the leakage of radioactive material. A 0.1-mrem radiation dose has 1 chance in 40 million of inducing a cancer, whereas one person in five dies from cancers due to other causes, so there is no more than 1 chance in 8 million that her cancer was connected with the leakage. Actually, it is surely less than that, because her cancer was not of a type normally induced by radiation. Moreover, 0.1 mrem was the maximum possible dose to any person; the dose received by this particular woman was probably very much lower.

As a result of this publicity and the public's extreme sensitivity to even the slightest radiation exposure from the nuclear industry, plus some honest desire on the part of government officials to improve security, a new set of regulations was formulated.[5] It requires that (1) the trench bottoms be well above any accumulating groundwater (i.e., the "water table") so as to exclude the possibility of the trenches filling with water; (2) surface covers be installed to minimize water passing through the trenches; (3) the waste be packed in such a way that the package maintains its size and shape even under heavy external pressures, when wet, or when subject to other potential adverse chemical and biological conditions; (4) packaging material be more substantial than cardboard or fiberboard; (5) there be essentially no liquid in the waste (excess water must be evaporated off); (6) empty spaces between waste packages be carefully filled; and so on.

Since the movement of water through the trenches can only be downward toward the water table, any radioactivity that escapes from the packages can only move downward until it reaches the water table, after which it can flow horizontally with the groundwater flow, which normally discharges into a river. This movement, as was explained in our discussion of high-level waste, must take many hundreds or thousands of years, because the water moves very slowly, and the radioactive materials are constantly being filtered out by the rock and soil.

In order to estimate the hazards from a low-level waste burial ground,[2] we must consider two possible routes low-level waste could take to get from the ground into the human stomach: (1) it could be picked up by plant roots and thereby get into food, and (2) it could be carried into a river and thereby get into water supplies. To evaluate the food pathway, we assume the unfavorable situation in which all of the waste escapes from its packaging and somehow becomes randomly distributed through the soil between the surface and the top of the water table. For natural materials, we know how much of each chemical element resides in this soil and how much resides in food, so we can calculate the probability per year for transfer from the soil into food. This probability then can be applied to the radioactive material of the same element, and of other elements chemically similar to it. When this probability is multiplied by the number of cancers that would result if all the low-level waste were ingested by people (calculated using the methods described in the Chapter 11 Appendix for the quantities of low-level waste generated by the nuclear industry), the product gives the number of deaths per year expected via the food pathway.

It is estimated that it would take something like 800 years for any of the radioactivity to reach a river. In order to estimate conservatively the hazards from the drinking water pathway, we therefore assume that all radioactivity remaining after 800 years reaches a river, and as in the case of high-level waste explained in Chapter 11, one part in 10,000 of it enters a human stomach.

Adding the effects of the food and drinking water pathways, we obtain the total number of deaths from the low-level waste generated by one large nuclear power plant in 1 year to be 0.0001 over the first 500 years and 0.0005 eventually.[2]

Siting of Low-Level Waste Burial Grounds

While we have shown that the hazards from low-level waste burial are very minimal, the public perception is quite the opposite. We have already noted how the burial grounds in Maxey Flats, Kentucky, and western New York State were shut down by adverse publicity. A third commercial burial ground near Sheffield, Illinois, was permanently closed when the space allocated in the original license became filled and an application for additional space was rejected. I was involved in the hearings leading to that decision. The setting had a strong "backwoods" flavor, and emotion ran rampant. No one seemed interested in my risk analyses and other quantitative information. The county supervisors decided not to increase the allocation.

That left three remaining burial grounds for commercial waste—at Barnwell, South Carolina, Beatty, Nevada, and Richland, Washington. Barnwell now accepts more than half of the total volume of waste, which contains more than three-fourths of the radioactivity. Richland accepts most of the remainder. With all the bad publicity, these sites became political liabilities to the states' governors, leading them to make threats that they might stop accepting waste from out of state. Politicians loudly proclaimed that it wasn't fair for their state to accept everyone else's waste. All three sites escalated their charges to customers constantly, nearly 10-fold over the past decade.

In order to resolve these problems, Congress passed legislation in 1980, amended in 1986, requiring each state to take responsibility for the low-level radioactive waste generated within its borders. The law contained a series of deadlines. By July 1, 1986, each state had to pass legislation specifying either that it would build a facility or that it had made arrangements for its waste to be sent to another state's facility; by January 1, 1988, each state was required to have legislation specifying the site selection procedure. Failure to meet any deadline meant that waste from that state must pay a hefty surcharge when sent for burial, and after January 1, 1993, the three states which now have burial grounds may refuse to accept waste from other states.

In response to this law, states formed compacts. For example, the Appalachian Compact consists of Pennsylvania, West Virginia, Delaware, and Maryland, with a disposal site in Pennsylvania. Present arrangements are to have sites in at least 13 states, with 6 of these scheduled to be operational by January 1, 1993. After that, states not affiliated with these six must make arrangements with one of them, or store their waste until their facility is available.

The law seems to be working reasonably well. There are pressures to extend deadlines, but so far the government has refused to do so. There is a great deal of activity as contractors go through the process of site selection. Local opposition is often vocal, and there is lots of negative publicity, but progress is being made.

TRANSURANIC WASTE

Although not found in nature, elements heavier than uranium, called *transuranics* (TRU), are produced in reactors and therefore become part of the waste, both high level and low level. These elements, like plutonium, americium, and neptunium, have special properties that make them more dangerous than most other radioactive materials if they get into the body. Since they are often relatively easily separated from other low-level wastes, it is government policy to do this where practical, and to dispose of these TRU wastes by deep burial somewhat similar to that planned for high-level waste. The first repository for this purpose, designated for waste from military programs, is now under construction in New Mexico. The amount of radioactivity in the TRU waste is many times less than in the high-level waste. Furthermore, elevated temperatures, which many consider to be the principal threat to waste security, are not a factor for TRU waste. Therefore, while there has been little study of commercial TRU waste disposal (there will be no commercial TRU waste until reprocessing is instituted), it seems clear that it represents less of a health hazard than high-level waste.

SUMMARY OF RESULTS

Table 1 lists the health effects from electricity generation that have been discussed in this and the previous chapter in terms of the number of deaths caused by one year's operation of a large power plant during the first 500 years, and eventually over multimillion-year time periods.[2] The minus sign

TABLE 1

SUMMARY OF THE NUMBER OF DEATHS CAUSED
BY THE WASTE GENERATED BY ONE LARGE POWER PLANT IN ONE YEAR,
OR BY THE EQUIVALENT AMOUNT OF ELECTRICAL ENERGY PRODUCTION

Source	Deaths caused	
	First 500 years	Eventually
Nuclear		
High-level waste	0.0001	0.018
Radon emissions	0.00	−420
Routine emissions (Kr-Xe,T,C-l4)	0.05	0.3
Low-level waste	0.0001	0.0004
Coal		
Air pollution	75	75
Radon emissions	0.11	30
Chemical carcinogens	0.5	70
Photovoltaics for solar energy		
Coal for materials	1.5	5
Cadmium sulfide	0.8	80

for radon emissions from nuclear power indicate that lives are saved rather than lost.

This table is the bottom line on the waste issue. It shows that, in quantitative terms, radioactive waste from nuclear power is very much less of a hazard than the chemical wastes, or even the radioactive wastes, from coal burning or solar energy. Almost every technology-based industry uses energy derived from coal and produces chemical wastes, and in nearly all cases, these are more harmful than the nuclear waste. This is true even ignoring the lives saved by mining uranium out of the ground; if the latter is included, nuclear waste considerations give a tremendous net saving of lives. By any standard of quantitative risk evaluation, the hazards from nuclear waste are not anything to worry about.

The problem is that no one seems to pay attention to quantitative risk analyses. There have been several books written (always by nontechnical authors) about the hazards of nuclear waste, without a mention of what the hazards are in quantitative terms. In one case the author interviewed me, at which time I went to great lengths in trying to explain this point, but he didn't seem to understand. When I explained what would happen if all the radioactive waste were ingested by people, he was busily taking notes, but when I tried to explain how small the probability is for an atom of buried waste to find its way into a human stomach, the note-taking stopped, and he showed impatience and eagerness to get on to other subjects. My impression was that he was writing a book to tell people about how horrible nuclear waste is, and his only interest was in gathering material to support that thesis.

THE REAL WASTE PROBLEM

The real waste problem has been waste of taxpayers' money spent to protect us from the imagined dangers of nuclear waste. One example is the handling of high-level waste from production of materials for nuclear bombs at the Savannah River plant in South Carolina.[6] This military high-level waste is considerably less radioactive than the commercial waste discussed in Chapter 11, so the DOE developed and carefully studied several alternative plans. One relatively cheap ($500 million) one was to pump the waste dissolved in water deep underground, whereas a much more expensive ($2.7 billion) one was to handle it like civilian high-level waste, converting it to glass and placing it in an engineered deep underground repository. The DOE study estimated that the first plan would eventually lead to eight fatal cancers spread over tens of thousands of years into the future (assuming that there will be no progress in fighting cancer). The DOE therefore decided to spend the extra $2.2 billion to save the eight future lives, a cost of $270 million per life saved. If this $2.2 billion were spent on cancer screening or highway safety, it could save 10,000 American lives in our generation. As one example, we could put air bags in all new cars manufactured this year and thereby save over a thousand lives per year for the next several years. Or we could install smoke alarms in every American home, saving 2,000 lives each year for some time to come.

But even these comparisons do not fully illustrate the absurdity of our unbalanced spending, because people living here many thousands of years in the future have no closer relationship to us than people now living in underdeveloped countries where millions of lives could be saved with this $2.2 billion. Some people have argued that it doesn't really help to save lives in countries suffering from chronic starvation because doing so increases their population and thus aggravates the problem. But that argument does not apply to areas where catastrophic 1-year famines have struck. For example,[7] during the 1970s, there were two such famines in Bangladesh that killed 427,000 and 330,000, a localized famine in India that killed 829,000, and one each in the African Sahel and Ethiopia that were responsible for 100,000 and 200,000 deaths, respectively. If we place such a low value on the lives of these peoples, how can we place such a high value on the lives of whoever happens to be living in this area many thousands of years in the future?

WEST VALLEY—THE ULTIMATE WASTE PROBLEM

The most flagrant waste of taxpayer dollars in the name of nuclear waste management is going on at West Valley, New York, about 30 miles south of Buffalo.[8] Since the West Valley problem has been widely publicized, it is worth describing in some detail. This was the site of the first commercial fuel-reprocessing plant, completed in 1966 and operated until 1972, when it was shut down for enlargement to increase its capacity. During the following few years, government safety requirements were substantially escalated, making the project uneconomical: the original cost of the plant was $32 million, and the initial estimated cost of the enlargement was $15 million, but it would have cost $600 million to meet the new requirements for protection against earthquakes. (All areas have some susceptibility to earthquakes, but it is minimal in the West Valley area.) It was therefore decided to abandon the operation, raising the question of what to do with the high-level waste stored in an underground tank.

The potential hazard was that the radioactive material might somehow leak out, get into the groundwater, and be carried with it into a nearby creek which runs into Lake Erie. Lake Erie drains into Lake Ontario and eventually into the St. Lawrence river, and the three of these are used as water supplies for millions of people. How dangerous would this be?

If all of the radioactive waste stored at West Valley were dissolved in Lake Erie now, and if it passed unhindered through the filters of water supply systems with no precautions being taken, we could expect 40,000 eventual fatalities to result. However, the radioactivity decreases with time by about a factor of 10 per century for the first few hundred years, so that if it were dumped into Lake Erie 400 years from now, only six fatalities would result; and if the dumping occurred more than 1,000 years in the future, there would probably not be a single fatality.

How likely would it be for wastes to get into Lake Erie in the near future? Let us suppose that all the containment features designed into the system failed, releasing all of the radioactive material into the soil. The nearest creek is several hundred feet away, and water soaking through the soil would take 10 to 100 years to traverse this distance. But the radioactive material would travel much more slowly; it would be effectively filtered out as the water passed through the soil and would consequently take 100 times longer—a total of at least a thousand years to reach the creek. We see that this alone gives a very high probability that the material will not get into the creek or lakes until its radioactivity is essentially gone.

But how likely is a release into the soil? The initial protection against this is the tank in which the waste is contained. It is basically one tank inside another, so that if the inner tank leaks, the radioactive material will still be contained by the outer tank and a warning about the situation will be given. In addition, there are three further barriers keeping it from getting into the soil. First, the tanks are in a concrete vault which should contain the liquid. Second, the concrete vaults are surrounded by gravel, and there are pipes installed to pump water out of this gravel. If the radioactivity managed to get into this region, it could still be pumped out through these pipes; there would be plenty of time—many weeks, at least—to do this. Third, the entire cavity is in a highly impermeable clay that would take a very long time for the liquid to penetrate before reaching the ordinary soil. There is still one last barrier worthy of mention; the water flow in the creek is sufficiently small that during the 10 or more years it would take the groundwater to reach it, a system could be set up for removing the radioactivity from the creek water.

Some perspective on the danger of leakage into the ground may be gained from considering a Russian program in which more than twice the radioactive content of the West Valley storage tank was pumped down a well into the ground. This was done as an experiment to study movement of the radioactivity through the ground with a view to using this method for large-scale high-level waste disposal. At last report the results were consistent with expectations and the plans were proceeding.

Up to this point we have been assuming that the radioactive materials are in solution in the waste storage tank, but actually 95% of them are in a solid sludge which is lying on the bottom of these tanks. This sludge would be much less likely to get out through a leak, to penetrate the concrete vault, and to be transported through the ground with groundwater; even if it were dumped directly into Lake Erie, most of it would settle to the bottom, and even if it got into city water supplies, it would very probably be removed by the filtration system. The consequences of release into Lake Erie that we have given earlier are therefore probably 10 times too high.

In summary, if there should be leakage from the tank, it would very probably be contained by the concrete vault. If it were not, it could be pumped out with the water which permeates the surrounding gravel. If this should fail, it would be contained for many years by the thick clay enclosing the entire cavity. When it did eventually get through to the surrounding soil, the movement of the radioactive materials would be sufficiently slow that they would decay to innocuous levels before reaching the creek. It would not be difficult to remove the radioactive material from the creek itself if this were necessary; if, as seems virtually certain, the material was delayed from reaching Lake Erie for at least 400 years, less than one fatality would be expected. If all else failed, any excess radioactivity in Lakes Erie and Ontario would be detected by routine monitoring operations, allowing precautions to be taken to protect public health.

But what if there were a violent earthquake? A structural analysis indicates that even the most violent earthquake believed possible in that area would not rupture the waste storage tanks. (Such an earthquake is expected only once in 16,000 years.) One might consider sabotage of the tank with explosives; but the tank is covered with an 8-foot thickness of clay which would be extremely hazardous to dig through unless elaborate protective measures were taken. If the tank were successfully ruptured, all of the other protective barriers would remain intact, so in all likelihood no harm would result. Saboteurs have many more inviting targets available if their aim is to take human lives. As an example, they could easily kill thousands of people by introducing a poison gas into the ventilation system of a large building.

A very large bomb dropped from an airplane could reach the waste and vaporize it: if this happened, several hundred fatalities would be expected, but far more people would be killed if this bomb were dropped on a city. These same considerations apply to a possible strike by a large meteorite or the development of a volcano through the area. These latter events are, of course, extremely improbable.

Up to this point we have ignored the effects of the radioactive materials permeating the soil in the event of a leak in the tank followed somehow by bypass of the concrete vault, the gravel pump-out system, and the thick clay lining. While in the soil, the radioactive materials could be picked up by plants and get into human food. How much of a hazard would this be? If all the radioactivity in the West Valley waste storage tank were now to become randomly distributed through the soil from the surface down to its present depth, if its behavior in soil is like that in average U.S. soil with the same percentage of land area used for farming, and if no protective action were taken, we would expect 30 fatalities. If the situation were postponed for 100 years, 3 fatalities would result, and if it were postponed for more than 180 years, we would not expect any. Our assumption here that the material becomes randomly distributed through the soil up to the surface is probably a very pessimistic one. Also, in the very unlikely event in which there could be a problem, it would easily be averted by checking food grown in the area for radiation and removing from the market any with excessive radioactivity.

In 1978, the DOE set about deciding what to do about this waste tank.[9] The simplest solution would be to pour cement mix into the tank to convert its contents into a large block of cement. This would eliminate any danger of leakage. The principal danger would then be that groundwater could somehow penetrate successively through the clay barrier, the concrete vault, and the stainless steel tank wall to dissolve away some of this cement. Each of these steps would require a very long time period. For example, although the sides of swimming pools and dams are cement, we note that they aren't noticeably leached away in many years even by the soaking in water to which they are exposed; moreover, groundwater contact is more like a dampness than a soaking. If the material did become dissolved in groundwater, all the barriers to getting into Lake Erie outlined above would still be in place and would have to be surmounted before any harm could be done. Even this remote danger could be removed by maintaining surveillance—periodically checking for water in the concrete vault and pumping it out if any should accumulate. The cost of converting to cement would be about $20 million, and a $15 million trust fund could easily provide all the surveillance one might desire for as long as anyone would want to maintain it.

If this were done, what would the expected health consequences be? I have tried to do risk analyses by assigning probabilities, and I find it difficult to obtain a credible estimate higher than 0.01 eventual deaths. It would be very easy to support numbers hundreds or thousands of times smaller.

However, this management option is not being taken. Instead the DOE has decided to remove the waste from the tank, convert it to glass, and bury it deep underground in accordance with plans for future commercial high-level waste. This program will cost about $1 billion. Spending $1 billion to avert 0.01 deaths corresponds to $100 billion per life saved! This is going on at a time when the same government is turning down projects that would save a life for every $100,000 spent! That is our real waste problem.

One last item deserves mention here—the radiation exposure to workers in executing the plans described above. It turns out that exposure is greater in the billion-dollar plan that was adopted than in the plan for conversion to cement, by an amount that would cause 0.02 deaths (i.e., a 2% chance of a single death) among the workers. Since this is more than 0.01 deaths to the public from the conversion to cement, the billion-dollar plan is actually more dangerous.

I have met the government officials who chose the billion-dollar plan, and have discussed these questions with them. They are intelligent people trying to do their jobs well. But they don't view saving lives as the relevant question. In their view, their jobs are to respond to public concern and political pressures. A few irrational zealots in the Buffalo area stirred up the public there with the cry "We want that dangerous waste out of our area." Why should any local people oppose them? Their congressional representatives took that message to Washington—what would they have to gain by doing otherwise? The DOE officials responded to that pressure by asking for the billion-dollar program. It wasn't hurting them; in fact, having a new billion-dollar program to administer is a feather in their caps. Congress was told that a billion dollars was needed to discharge the government's responsibility in protecting the public from this dangerous waste—how could it fail to respond?

That is how a few people with little knowledge or understanding of the problem induced the United States Government to pour a billion dollars "down a rathole." I watched every step of the process as it went off as smooth as glass. And the perpetrators of this mess have become local heroes to boot.

LEAKING WASTE STORAGE TANKS

No discussion of nuclear waste problems could be complete without including leaking waste storage tanks. There was a great deal of publicity about them and their hazards, especially in the early 1980s.

When high-level waste is first isolated in a chemical-reprocessing plant, it is stored in underground tanks for a few years before being converted into glass. One of these tanks might handle the waste accumulating from 50 power plants over several years, a very substantial quantity of radioactivity. This raises the question of the dangers from possible leaks in the tanks.

This question arose very early in the history of nuclear energy. During World War II, the Hanford Laboratory was established in the desert of central Washington State to produce bomb material in reactors and separate it in chemical-reprocessing plants, leaving the waste in underground storage tanks. These tanks were made of ordinary steel, which was known to be readily corroded by the waste solution. It was therefore assumed that the tanks would eventually leak, releasing the radioactive waste into the soil. The thinking at that time was that this would be an acceptable situation—if the radioactive material was eventually to be buried underground anyway, why not let some of it get there through leaks in storage tanks? However, as public concern about radiation escalated over the years, this procedure became less acceptable, and in the 1960s the technology was changed. The new tanks were thereafter constructed of stainless steel, which is much less easily corroded, and facilities were included to keep close track of any corrosion that might occur. But more important, the new tanks were constructed with double walls. Thus, if the inner wall developed a leak, the liquid would fill the space between the walls, thereby signaling the existence of the leak. In the meantime, the liquid would still be contained by the outer tank wall, which would leave plenty of time to pump the contents into a spare tank.

These new tanks, which have been used in several locations for many years, including the West Valley tank described above, have had no problems. But some of the old single-wall tanks at Hanford have developed leaks, as was to be expected. On such occasions, the practice has been to pump the contents into a spare tank, but on one occasion in particular, a leak went undetected for 7 weeks, resulting in discharge of a substantial quantity of radioactivity into the soil. Although there has been extensive adverse publicity from the incident—and there was irresponsible negligence involved—there have been no health consequences. Moreover, it is most difficult to imagine how there can ever be any, due to the following considerations.

All of the significant radioactive materials are now absorbed in the soil within a few feet of the tank, still 150 feet above the water table. As rainwater occasionally percolates down through the soil, some of this material may be expected eventually to reach the water table after several hundred years. Only then can it move horizontally, toward the Columbia River, about 10 miles away. However, groundwater flows very slowly in that region, requiring 800 years to cover that 10-mile distance. The only radioactive material involved in the leak that will last more than a few hundred years is plutonium, which survives to some extent for a few hundred thousand years, but plutonium in groundwater is constantly filtered out by the rock and hence travels 10,000 times more slowly than the water. It would therefore take 200,000 years to reach the river, by which time it would have decayed away.

However, even if all of the plutonium involved in the leak (about 50 grams) were dumped into the river now, there is less than a 1% chance that even a single human health effect would result. While the analysis given here contains no controversial elements and has never, to the best of my knowledge, been scientifically questioned, I have frequently heard word-of-mouth claims of detrimental health effects, both present and future, from the leaking tanks at Hanford. Rarely are such effects not at least hinted at in popular books and magazine articles about radioactive waste.

Some readers may be surprised by the above statement that 50 grams

(1.8 ounces) of plutonium dumped in a river is so harmless. Part of the reason for this is that only one atom in 10,000, or 0.005 grams, can be expected eventually to enter human stomachs (see Chapter 11 Appendix). Another part of the reason is that when plutonium does enter a stomach, 99.99% of it is excreted within a few days, so it has little opportunity to irradiate the vital body organs. The stories one hears about the high toxicity of plutonium are all based on inhaling it into the lungs, rather than ingesting it with food or water.

WASTE TRANSPORT — WHEN RADIOACTIVITY ENCOUNTERS THE PUBLIC

Spent fuel must be shipped from reactors to reprocessing plants, and the high-level waste derived from it must be shipped to a repository for burial; therein lie substantial potential hazards. After all, inside plants the radioactivity is remote from the public, but in transport close proximity is unavoidable. Moreover, accidents are inherently frequent in transportation — half of all accidental deaths occur during the small percent of the time we spend traveling.

We will confine our attention to the shipping of spent fuel, since that is where there has been the most experience; other high-level waste shipments will be handled analogously. One general safety measure is to delay shipping as long as possible to allow short-lived radioactivity to decay away. For spent fuel, the minimal delay has been 6 months. But the most important safety device is the cask in which the spent fuel is shipped. It typically costs a few million dollars, and one can well imagine that a great deal of protection can be bought for that kind of money.

These casks have been crashed into solid walls at 80 miles per hour, and hit by railroad locomotives traveling at similar speeds, without any release of their contents.[10] These and similar tests have been followed by engulfment in gasoline fires for 30 minutes and submersion in water for 8 hours, still without damage to the contents. In actual practice, these casks have been used to carry spent fuel all over the country for more than 40 years. Railroad cars and trucks carrying them have been involved in all sorts of accidents, as might be expected. Drivers have been killed; casks have been hurled to the ground; but no radioactivity has ever been released, and no member of the public has been exposed to radiation as a consequence of such accidents.

If we try to dream up situations that could lead to serious public health consequences, we are limited by the fact that nearly all of the radioactivity is solid material, unable to leak out like a liquid or a gas. There is no simple mechanism for spreading it over a large area even if it did get out of the cask. In almost any conceivable situation, significant radiation exposure would be limited to people who linger for several minutes in the immediate vicinity of the accident; hence the number of people exposed would be relatively small.

When all relevant factors are included in an analysis, studies indicate that with a very full nuclear power program there would eventually be one death every few thousand years in the United States resulting from radioactivity releases in spent fuel transport accidents.[11] Of course, there would be many times that number of deaths from the normal consequences of these accidents.

Some people have posed the problem of terrorists blowing up a spent fuel cask while it is being transported through a city.[12] To study this, Sandia National Laboratory carried out tests with high-explosive devices and lots of instruments to help determine just what was happening. Their conclusion was that even if several hundred pounds of high explosives were used on a cask traveling through downtown Manhattan at noon on a week day, the total expected number of eventual deaths would be 0.2; that is, there is only a 20% chance that there would be a single death from radiation-induced cancer.

A more important effect is the slight radiation exposure to passers-by as trucks carrying radioactive waste travel down highways. Some gamma rays emitted by the radioactive material can penetrate the walls of the cask to reach surrounding areas. (Due to weight limitations, it is not practical to have as thick a shield wall on a truck as is used around reactors.) Even with a full nuclear power program, however, exposures to individuals would be a tiny fraction of 1 mrem. The effects would add up to one death every few centuries in the United States.[11]

Some perspective on these results can be obtained by comparing them with impacts of transport connected with other energy technologies. Gasoline truck accidents kill about 100 Americans each year and injure 8 times that number. It has been estimated that coal-carrying trains kill about 1,000 members of the public each year. Clearly, the hazards in shipment of radioactive waste are among the least of our energy-related transportation problems.

In spite of the very long and perfect record of spent fuel transport and its extremely small health effects, a great deal of public fear has been generated. Many municipalities, ranging from New York City down to tiny hamlets, have consequently passed laws prohibiting spent fuel transport through their boundaries. It does not take many such restrictions to cause extreme difficulty in laying out shipping routes. A New York City ordinance, for example, prevents any rail or truck shipments from Long Island to other parts of the United States.

A RADIOACTIVE WASTE ACCIDENT IN THE SOVIET UNION

In the late 1970s, there was a great deal of publicity about a disastrous accident involving high-level nuclear waste in the Soviet Union. The incident reportedly took place near the Kyshtym nuclear weapons complex in the southern Ural Mountains during the winter of 1957-1958. First reports[13] were from Z. A. Medvedev, a Soviet scientist now living in England, who pieced together information from rumors circulating among Russian scientists and from radiation contamination studies reported in the Russian literature. A second Soviet scientist, now living in Israel, reported having driven through the area in 1960 and observing that the area was not inhabited; his driver told him that there had been a large explosion there. Medvedev theorized that this situation had resulted from a nuclear explosion of radioactive waste, and according to rumors he had heard, there had been many casualties and serious radioactive contamination over an area of several thousand square miles.

Following these reports, extensive studies were carried out at Oak Ridge National Laboratory,[14] and later at Los Alamos National Laboratory,[15] using information culled from a thorough search of the Russian literature and U.S. Government (CIA) sources. It was concluded that the incident resulted from some very careless handling of radioactive waste in the Russian haste to build a nuclear bomb. It was deduced that there had been a chemical explosion due to use of a reprocessing technology that is very efficient but was rejected for use in the United States because of the danger of such an explosion. Both groups concluded that it could not have been caused by a nuclear explosion of buried radioactive waste as had been conjectured by Medvedev, because the ratio of the quantities of various radioactive materials was grossly different from that produced by a nuclear explosion. Moreover, a large nuclear explosion requires that very special materials be brought together very rapidly, in a tiny fraction of a second, and that very little water be present, whereas waste is normally dissolved in water, and the movement of materials is very slow. The U.S. researchers became quite convinced that the area contaminated was much smaller than Medvedev had estimated—rumor mills often exaggerate such figures. None of the accidents that seemed even remotely possible to the American researchers could have administered radiation doses approaching a lethal range. It is therefore highly dubious whether there were any casualties from radiation. Moreover, there is nothing in the Russian scientific literature on medical effects of radiation to suggest that any had occurred in this incident.

The most definite conclusion of both U.S. studies was that no such accident could possibly occur with American waste-handling procedures — answering that question was the prime reason for these studies. None of the ingredients that could lead to either a nuclear or a chemical explosion is present in the U.S. technology.

As a result of their new *glasnost* policy, the Soviet Union released a report[16] on the incident in 1989. It confirmed that on September 29, 1957, there had been a chemical explosion of the type conjectured by the U.S. studies in a high-level waste storage tank. The area treated as contaminated was about 400 square miles. About 600 people were evacuated in the first few weeks, and the number eventually grew to 10,000. There were no casualties from the explosion, and medical and epidemiological studies over the past 30 years indicate no excess mortality rates or excess disease incidence among those exposed, and no genetic effects in their progeny. Nearly all of the land has now been restored for use, mostly as farms. The report emphasizes that experience and data from the Kyshtym incident have been very useful in dealing with consequences of the Chernobyl accident.

Opponents of nuclear power in the United States used the Kyshtym incident as an indication of the hazards of radioactive waste. However, it is important to emphasize that the technology that led to the chemical explosion in Kyshtym has never been used in the United States because that danger was recognized. Such an accident, therefore, most definitely cannot happen here.

SUMMARY

This chapter and the previous one have presented detailed scientific analyses of the radioactive waste problem. The principal results of these analyses are contained in Table 1, with additional information in the section on waste transport. They show that the hazards of nuclear waste are thousands of times lower than those of wastes from other technologies. However, the American public has been badly misinformed about these dangers and has become deeply concerned about them. As a result, billions of dollars are being wasted, but fortunately, even with all this unnecessary spending, the cost of radioactive waste management increases the price of nuclear electricity by only about one percent.

Chapter 13 / PLUTONIUM AND BOMBS

The very existence of plutonium is often viewed as the work of the devil.* As the most important ingredient in nuclear bombs, it may someday be responsible for killing untold millions of people, although there are substitutes for it in that role if it did not exist. If it gets into the human body, it is highly toxic. On the other hand, its existence is the only guarantee we have that this world can obtain all the energy it will ever need *forever* at a reasonable price. In fact, I am personally convinced that citizens of the distant future will look upon it as one of God's greatest gifts to humanity. Between these extremes of good and evil is the fact that if our nuclear power program continues to be run as it is today, the existence of plutonium will have no relevance to it except as a factor in technical calculations.

Clearly, there are several different stories to tell about plutonium. We

*It is sometimes said that it was named with that in mind, but it was actually named for the planet Pluto.

will start with the future benefits, then discuss the weapons connection, and conclude with the toxicity question.

FUEL OF THE FUTURE

As uranium occurs in nature, there are two types, U-235 and U-238, and only the former, which is less than 1% of the mixture, can be burned (i.e., undergo fission) to produce energy. Thus, present-day power reactors burn less than 1% of the uranium that is mined to produce their fuel. This sounds wasteful but it makes sense economically, because the cost of the raw uranium at its current price represents only 5% of the cost of nuclear electricity (see Chapter 13 Appendix). However, there is only a limited amount of ore from which uranium can be produced at anywhere near the current price, perhaps enough to provide lifetime supplies of the fuel needed by all nuclear power plants built up to the year 2025. Beyond that, uranium prices would escalate rapidly, doubling the cost of nuclear electricity within several decades.

Fortunately, there is a solution to this problem. The fuel for present-day American power plants is a mixture of U-238 and U-235. As the reactor operates, some of the U-238, which cannot burn, is converted into plutonium. This plutonium can undergo fission and thus serve as a nuclear fuel. In fact, some of it is burned while the fuel is in the reactor, enough to account for one-third of the reactor's total energy production. But some of it remains in the spent fuel from which it can be extracted by chemical reprocessing. This plutonium could be burned in our present power reactors, but an alternative is to use it in another type of reactor, the *breeder,* whose fuel is a mixture of plutonium and uranium (U-238). Much more of the U-238 in the breeder is converted to plutonium than in our present reactors, more than enough to replace all of the plutonium that is burned. Thus, a breeder reactor not only generates electricity, but it produces its own plutonium fuel with extra to spare. It only consumes U-238, which is the 99 + % of natural uranium that cannot be burned directly; therefore, it provides a method for indirectly burning this U-238. With it, nearly all of the uranium, not less than one percent as in present type reactors, is eventually burned to produce energy. About a hundred times as much energy is thus derived from the same initial quantity. That means that instead of lasting only for about 50 years, our uranium supply will last for thousands of years. As a bonus, the environmental and health problems from uranium mining and mill tailings will be reduced a hundredfold. In fact, all uranium mining could be stopped for about 200 years while we use up the supply of U-238 that has already been mined and is now in storage.

Deriving 100 times as much energy from the same amount of uranium fuel means that the raw fuel cost per kilowatt-hour of electricity produced is reduced correspondingly. In fact, the fuel costs per unit of useful energy generated in a breeder reactor are equivalent to those of buying gasoline at a price of 40 gallons for a penny! (See Chapter 13 Appendix.) Instead of contributing 5% to the price of electricity as in present-type reactors, the uranium cost then contributes only 0.05% in a breeder reactor. If supplies should run short, we can therefore afford to use uranium that is 20 times more expensive, for even that would raise the cost of electricity by only $(20 \times .05 =)$ 1%. How much uranium is available at that price?

The answer is effectively *infinite* because it includes uranium separated out of seawater.[1] The world's oceans contain 5 billion tons of uranium, enough to supply all the world's electricity through breeder reactors for several million years. But in addition, rivers are constantly dissolving uranium out of rock and carrying it into the oceans, renewing the oceans' supply at a rate sufficient to provide 25 times the world's present total electricity usage.[2] In fact, breeder reactors operating on uranium extracted from the oceans could produce all the energy humankind will ever need* without the cost of electricity increasing by even 1% due to raw fuel costs.

The fact that raw fuel costs are so low does not mean that electricity from breeder reactors is very cheap. The technology is rather sophisticated and complex, involving extensive handling of a molten metal (liquid sodium) that reacts violently if it comes in contact with water or air. Largely as a result of the safety precautions required by this problem, the cost of electricity from the breeder will be substantially higher at today's uranium prices than that from reactors now in use.[3] Nevertheless, France, England, and the Soviet Union have continued with developing breeder reactors, and several other countries, including Germany and Japan, are involved to a lesser degree. The American program was at the forefront 20 years ago, but it became a political football and is now essentially dead.

On the surface, the opposition to the U.S. breeder reactor is based on the fact that uranium supplies are plentiful and cheap, leaving little incentive for an expensive development program at this time (less expensive research is continuing, most notably in a test reactor at the Hanford site in Washington State). Why, then, have other countries continued to press on with their development programs? First, even if development goes forward at the

---

*The earth is expected to last for about 5 billion years before it becomes a molten mass due to changes in the sun. The uranium supply is adequate for that time period.

hoped-for pace, it will be many years before the first commercial breeder can become operational and many more before its use would become widespread; it is better to start up any new technology slowly, allowing the "bugs" to be worked out before a large number of plants are built. Second, we are not that certain about our uranium resources; they may be substantially below current estimates. Having the breeder reactor ready would be a cheap insurance policy against that eventuality, or against any sharp increase in uranium prices for whatever the reason. And third, the breeder reactor development program has substantial momentum, with lots of scientists, engineers, and technicians deeply involved. It is much more efficient to carry the program to completion now than to stop it, allow these people to become scattered, and then start over with a new team of personnel later.

Not far beneath the surface, there is substantial opposition to the breeder because of distaste for plutonium and general opposition to nuclear power. There are also some fears about the safety of breeder reactors, but experts on that subject (of which I am not one) maintain that they are extremely safe, and even safer than present reactors.[3-8] They have the important safety advantage of operating at normal pressure rather than at very high pressure, as is the case for present reactors. There are therefore no forces tending to enlarge cracks or to blow the coolant out of the reactor (this is the "blowdown" discussed in Chapter 6).

A key part of the breeder reactor cycle is the reprocessing of spent fuel to retrieve the plutonium. In fact, this must be done with the spent fuel from present reactors in order to obtain the plutonium necessary to fuel the first generation of breeder reactors. As long as there is no reprocessing, the plutonium occurs only in spent fuel, where it is so highly dilute (one-half of 1% of the total) that it is unusable for any of the purposes usually discussed. Moreover, spent fuel is so highly radioactive (independently of its plutonium content) that it can only be handled by large and expensive remotely controlled equipment. It therefore cannot be readily stolen or used under clandestine conditions. Without reprocessing, there is no use for plutonium for good or evil.

It should also be recognized that plutonium plays only a minor role in waste disposal problems, and a negligible role in reactor accident scenarios. Thus, as long as there is no reprocessing, which is the present status in the United States commercial nuclear power program, plutonium issues have no direct relevance to the acceptability of nuclear power.

However, it is my personal viewpoint that it is *immoral* to use nuclear power without reprocessing spent fuel. If we were simply to irretrievably bury it, we would consume all the rich uranium ores within about 50 years.

This would deny future citizens the opportunity of setting up the breeder cycle, the only reasonably low-cost source of energy for the future of which we can be certain. By such action, our generation might well go down in history as the one that denied humankind the benefits of cheap energy for millions of years, a fitting reason to be eternally cursed. On the other hand, if we develop the breeder reactor, we may go down in history as the generation that solved the world's energy problems for all time. Future generations might well remember and bless us for millions of years.

Unfortunately, the people in control are not worried about the long-range future of mankind. People in the nuclear power industry are concerned principally about the next 30 or 40 years, and politicians rarely extend their considerations even that far into the future. Whether or not we do reprocessing will have little impact over these time periods; thus the prospects for early reprocessing are questionable.

The situation was very different only a few short years ago. A large reprocessing plant capable of servicing most of the power plants now operating in the United States was constructed near Barnwell, South Carolina, by a consortium of chemical companies. The main part of the plant, costing $250 million, was completed in 1976, but two add-ons that would have cost about $130 million were delayed by government indecision. Since the add-ons would not be needed for several years, it was expected that the main part of the plant could be put into immediate operation.

At that critical point, the U.S. Government decreed an indefinite deferral of commercial reprocessing. The reason for the decree involved our national policy on discouraging proliferation of nuclear weapons, which will be discussed later in this chapter, but from the viewpoint of the plant owners, it was a disaster. They had been strongly encouraged to build the plant by government agencies—for example, federally owned land was made available to them for purchase—and every stage of the planning was done in close consultation with those agencies. They had scrupulously fulfilled their end of the bargain, laying out a large sum of money, and now they were left with a plant earning no income.

By the time the Reagan Administration withdrew the decree forbidding reprocessing 5 years later, the owners had lost heart in the project and were unwilling to provide the money, now increased to over $200 million, to provide the add-ons. The Barnwell plant was abandoned. It is generally recognized that there will be no commercial reprocessing in the United States unless the government provides assurances that money invested would be compensated if the project were again terminated by political decree, and guarantees to purchase the plutonium it produces. The latter requirement is necessary because the Barnwell plant was originally built with the understanding that utilities could purchase the plutonium to fuel present reactors, but the government has not taken action to allow this and probably will never do so. It is now widely agreed that it would be better to save the plutonium for breeder reactors. Since there are no commercial breeder reactors in the United States and will not be any for many years, this leaves the government as the only customer for the plutonium from a reprocessing plant.

Aside from the idealistic considerations of providing energy for future generations, an additional driving force behind getting reprocessing plants into operation is their contribution to waste management. Power plants are having difficulty in storing all of the spent fuel they are discharging; reprocessing gives them an outlet for it. Furthermore, the amount of material to be buried is very much reduced if the uranium is removed in reprocessing. There is also considerably more security in burying high-level waste converted to glass and sealed inside a corrosion-resistant casing, than in burying unreprocessed spent fuel encased in asphalt or some similar material.

On the other hand, there has been strong opposition to reprocessing. There have been well publicized attacks on its environmental acceptability, ignoring the contrary evidence in the scientific literature in favor of "analyses" by "environmental groups" tailored to reach the desired conclusion. There were widely publicized economic analyses of unspecified origin claiming that reprocessing was a money-losing proposition, even when the real professionals in the business considered it to be economically advantageous.[9] There was a considerable amount of publicity for a paper issued by the DOE claiming that the Barnwell plant was technically flawed,[10] but it turned out the paper was by a scientist with little experience in the field who had never visited the plant and was confused over differences between reprocessing fuel from present power reactors and breeder reactors; the paper had accidently slipped through the DOE reviewing process and was disavowed and strongly critiqued by the head of the division that had issued it.[11]

A major part of this opposition to reprocessing came from those opposed to nuclear power in general for political and philosophical reasons. They realized that it was too late to stop the present generation of reactors, but if they could stop reprocessing, nuclear power could have no long-term future. However, the most important opposition to reprocessing came from its possible connection to nuclear weapons. If there is a connection between nuclear electricity and nuclear explosives, reprocessing is the bottleneck through which it must pass. We now turn to a discussion of that matter.

PROLIFERATION OF NUCLEAR WEAPONS

Everyone agrees that nuclear weapons can have very, very horrible effects and that it is exceedingly important to avert their use against human targets. One positive step in this direction is to minimize the number of nations that have them available for use—that is, to avoid the proliferation of nuclear weapons. To what extent do nuclear power programs frustrate that goal?

If a nation has a nuclear power reactor and a reprocessing plant, it could reprocess the spent fuel from the reactor to obtain plutonium, and then use that plutonium to make bombs. On the other hand, there are much better ways for nations to obtain nuclear weapons. There are two* practical fuels for nuclear fission bombs: U-235, which occurs in nature as less than 1% of normal uranium, from which it must be removed by a process known as "isotope separation," and plutonium, which can be produced in nuclear reactors and converted into usable form through reprocessing. Either method can produce effective bombs, although the best bombs use a combination of both. Both the isotope separation and the reactor-reprocessing methods are used by all five nations known to have nuclear weapons, the United States, the Soviet Union, Great Britain, France, and China. (India has also exploded a nuclear device but claims that it was for nonweapons purposes.)

However, there is a subtle aspect to producing plutonium by the reactor-reprocessing method, and to explain it we will divert briefly to review our Chapter 7 discussion of how a plutonium bomb works. There are two stages in its operation: first, there is an implosion in which the plutonium is blown together and powerfully compressed by chemical explosives that surround it, and then there is the explosion in which neutrons are introduced to start a rapidly escalating chain reaction of fission processes that release an enormous amount of energy very rapidly to blow the system apart. All of this takes place within a millionth of a second, and the timing must be precise— if the explosion phase starts much before the implosion process is completed, the power of the bomb is greatly reduced. In fact, one of the principal methods that has been considered for defending against nuclear bombs is to

*There is a third possible fuel, uranium-233, but considering it would unduly complicate the discussion here.

shower them with neutrons to start the explosion early in the implosion process, thereby causing the bomb to fizzle. For a bomb to work properly, it is important that no neutrons come upon the scene until the implosion process approaches completion.

Plutonium fuel, Pu-239, is produced in a reactor from U-238, but if it remains in the reactor it may be converted into Pu-240, which happens to be a prolific emitter of neutrons. In a U.S. power plant, the fuel typically remains in the reactor for 3 years, as a consequence of which something like 30% of the plutonium produced comes out as Pu-240. If this material is used in a bomb, the Pu-240 produces a steady shower of 2 million neutrons per second,[12] which on an *average* would reduce the power of the explosion tenfold, but might cause a much worse fizzle. In short, a bomb made of this material, known as "reactor-grade plutonium," has a relatively low explosive power and is highly unreliable. It is also far more difficult to design and construct.

A much better bomb fuel is "weapons-grade plutonium," produced by leaving the material in a reactor for only about 30 days. This reduces the amount of Pu-240 and hence the number of neutrons showering the bomb by tenfold.

One might consider trying to use a U.S.-type power reactor to produce weapons-grade plutonium by removing the fuel for reprocessing every 30 days, but this would be highly impractical because fuel removal requires about a 30-day shutdown. Moreover, the fuel for these power reactors is very expensive to fabricate because it must operate in a very compact geometry at high temperature and pressure to produce the high-temperature, high-pressure steam needed to generate electricity.

It is much more practical to build a separate plutonium *production reactor* designed not to generate electricity but rather to provide easy and rapid fuel removal in a spread-out geometry with fuel that is cheap to fabricate because it operates at low temperature and normal pressure. Moreover, it can use natural uranium rather than the very expensive enriched uranium needed in power reactors. For a given quantity of fissile material, the former contains 4 times as much of the U-238 from which plutonium is made, hence producing 4 times as much plutonium. A plutonium production reactor costs less than one-tenth as much as a nuclear power plant[13] and could be designed and built much more rapidly. All of the plutonium for all existing military bombs has been produced in this type of reactor except in the Soviet Union where a compromise design allowing both electricity generation and plutonium production is employed (see Chapter 7).

Another alternative would be to use a *research reactor,* designed to provide radiation for research applications* rather than to generate electricity. At least 45 nations now have research reactors, and in at least 25 of these there is a capability of producing enough plutonium to make one or more bombs every 2 years. Research reactors are usually designed with lots of flexibility and space, so it would not be difficult to use them for plutonium production.

A plant for generating nuclear electricity is by necessity large and highly complex, with most of the size and complexity due to reactor operation at a very high temperature and pressure, the production and handling of steam, and the equipment for generation and distribution of electricity. It would be impossible to keep construction or operation of such a plant secret. Moreover, only a very few of the most technologically advanced nations are capable of constructing one. No nation with this capability would provide one for a foreign country without requiring elaborate international inspection to assure that its plutonium is not misused. A production or research reactor, on the other hand, can be small and unobtrusive. It has no high pressure or temperature, no steam, and no electricity generation or distribution equipment. Almost any nation has, or could easily acquire, the capability of constructing one, and it probably could carry out the entire project in secret. There would be no compulsion to submit to outside inspection.

In view of the above considerations, it would be completely illogical for a nation bent on making nuclear weapons to obtain a power reactor for that purpose. It would be much cheaper, faster, and easier to obtain a plutonium production reactor; the plutonium it produces would make much more powerful and reliable bombs with much less effort and expense.

The only reasonable scenario in which U.S.-type power reactors might be used is if a nation decided it needs nuclear weapons *in a hurry.* In such a situation, 1 or 2 years could be saved if a power reactor were available and a production or large research reactor were not.[13] However, nearly all nations that have a power reactor also have research reactors. Moreover, it would be most unusual for this time saving to be worth the sacrifice in weapons quality.

---

*Perhaps the most important application of research reactors is producing radioactive isotopes for use in medicine, agriculture, industry, education, etc. Another is to provide neutron beams for studying properties of materials ranging from metals to biological molecules. They provide the most sensitive means available for measuring tiny quantities of material such as the gunpowder residue on a criminal's hand from firing a gun, or the mercury in fish. There are many other applications of research reactors ranging from basic physics and chemistry to practical engineering and biology.

But obtaining plutonium is not the only way to get nuclear weapons. The other principal method is to develop *isotope separation* capability. Nine nations now have facilities for isotope separation,[13] and others would have little difficulty in acquiring it. A plant for this purpose, costing $20-200 million, could provide the fuel for 2-20 bombs per year and could be constructed and put into operation in 3-5 years.[13] The product material would be very easy to convert into excellent bombs, much easier than making a plutonium bomb even with weapons-grade plutonium.

This assessment is based on present technology, but several new, simpler, and cheaper technologies for isotope separation are under development and will soon be available. They will make the isotope separation route to nuclear weapons even more attractive. There are also new technologies under consideration for producing plutonium without reactors, which may make that route more attractive.

The way I like to explain the problem of nuclear weapons proliferation is to consider three roads to that destination: (1) isotope separation, (2) plutonium production with research or production reactors, and (3) plutonium production in U.S.-type power plants, with (2) and (3) requiring reprocessing. The first two roads are much more attractive than the third from various standpoints; they are like super highways, while the third is like a twisting back country road. In this analogy, how important is it to block off the third road while leaving the first two wide open? The link between nuclear power and proliferation of nuclear weapons is a weak and largely insignificant one.*

But that is certainly not the impression the public has received. The great majority of stories about nuclear weapons proliferation involves nuclear power plants. They generally give the impression that without nuclear power there would be no proliferation problem. They rarely differentiate between a power reactor and other types more suitable for making bombs. I believe most Americans think that the Iraqi reactor destroyed by an Israeli air raid was a nuclear power plant, when in fact it was a large research reactor.

Even though nuclear power plants are only a minor source of weapons proliferation, nobody is saying that elaborate precautions should not be taken to see that the plutonium in them is not used for that purpose. The

---

*There would be a somewhat stronger link — but still not a very strong one — with breeder reactors. However, it would not be difficult to limit breeder reactors to nations in which there is not an issue of weapons proliferation. In any case, no such nation will have a breeder reactor for the next several decades.

programs for dealing with that problem are known as "safeguards." They are administered by the International Atomic Energy Agency (IAEA) based in Vienna. The IAEA has teams of inspectors trained and equipped to detect diversion of plutonium. In nations subject to safeguards programs, the IAEA has ready access to all nuclear power plants, reprocessing plants, and other facilities involved in handling plutonium. (The principal other facility would be for fabricating plutonium fuel for use in breeder reactors or perhaps in present reactors if the price of uranium should decrease sharply.) There has been an impressive development in techniques and equipment for carrying out these inspections. For example, the Barnwell reprocessing plant developed an automatic computer-controlled system that gives a warning in less than an hour if any plutonium in the plant should not be where it is supposed to be. With such measures and IAEA inspectors on the scene or making unannounced visits, it would be very difficult for a nation to divert plutonium from its nuclear power program without the rest of the world knowing about it long before the material could be converted into bombs.

These safeguards would be much easier to circumvent with a production or research reactor or with an isotope separation plant. These are much smaller operations with far less support needed from foreign suppliers; it would not be difficult to build them clandestinely. The IAEA safeguards system thus does much more to block off the twisting back country road than the super highways.

NONPROLIFERATION POLITICS

One would have thought that these safeguards would be enough attention paid to the back country road, but the Carter Administration saw fit to go a step further. It decided to try to prevent the acquisition of reprocessing technology by nonnuclear weapons nations. As you may recall, reprocessing is a bottleneck that must be passed if nuclear power plants are to be used to make bomb materials; thus the goal of the government was, in principle, a desirable one. However, the method for implementing it was disastrous.

At that time (1977), Germany was completing a deal to set up a reprocessing plant in Brazil, Japan was building a plant, and France was negotiating the sale of plants to Pakistan and Korea. The Carter goal was to stop these activities through moral and political pressure. To set the moral tone for this effort—essentially to "show that our heart is in the right place"—he decided to defer indefinitely the reprocessing of commercial nuclear fuel in the United States.* This was the move that prevented the Barnwell plant from operating.

There were several problems with this approach. One was that the U.S. Government continued to do reprocessing in its military applications program, which was something of a dilution of the high moral tone being advertised. Another was that Germany had just won the Brazilian contract after stiff bidding competition with U.S. firms. The Germans therefore interpreted the U.S. initiative as sour grapes over the loss of business. But a much bigger problem arose from the political pressure used: the United States delayed and threatened to stop shipments of nuclear fuel to nations that would not cooperate.

American manufacturers had built up a thriving export business of selling reactors to countries all over the world. Part of the deals was a guaranteed future supply of fuel for the reactors; this meant U.S. Government participation in the contracts, because it possessed the only large-scale facilities for isotopic enrichment of uranium. These sales contracts had no clauses allowing interruption of the fuel supply—no one would spend hundreds of millions of dollars for a power plant without a guaranteed fuel supply—so the delays and threatened withholding of shipments by the United States represented a direct and illegal breach of contract. Even nations with no interest in reprocessing were deeply upset by the very principle of this action. I remember sitting in a frenzied session of a meeting in Switzerland on this subject. The session was in German, for the benefit of Swiss journalists, and I did not understand much of it, but I kept hearing the word "nonproliferationpolitik" accompanied by expressions of intense anger and banging on the table. At one point Yugoslavia, which purchased a Westinghouse reactor, was close to breaking off diplomatic relations with the United States over this issue.

Not only was withholding fuel shipments a breach of contract, but it was a violation of the International Treaty on Nonproliferation of Nuclear Weapons. That treaty states that a nonweapons nation that signs the treaty is entitled to a secure and uninterrupted supply of fuel for its power reactors. Thus the United States became the first nation to violate that treaty, which is the most important safeguard the world has against proliferation. Incidentally, this furor in Europe, Asia, and South America over the Carter initiative received virtually no media coverage in this country.

---

*Actually, Mr. Carter raised the issue in the 1976 presidential election campaign, and in response, President Ford called a temporary halt to reprocessing to allow time for review a few days before the election. The Carter Administration then turned it into a long-term instrument of policy.

But the worst problem with the Carter initiative was that it failed to achieve much in the way of results. The United States had enough political leverage over South Korea to force that country to cancel its purchase of a reprocessing plant. France cancelled its sale to Pakistan, probably in recognition of the fact that Pakistan had expressed ambitions for building nuclear weapons, but perhaps also partly as a result of American political pressure. However, the German deal with Brazil was not cancelled in spite of constant political pressure, including several face-to-face meetings between President Carter and German Chancellor Schmidt. The Japanese reprocessing plant was completed and started up. No other reprocessing activity anywhere in the world except in the United States was stopped by the Carter initiative.

While the Carter initiative had little impact on the international proliferation problem, it did have two very important negative effects in this country: it prevented the start-up of the Barnwell plant, as discussed earlier, which has had a long-lasting devastating effect on commercial reprocessing in the United States; and it has completely ruined the U.S. reactor export business. Several nuclear power plants are purchased by foreign countries every year, and at one time, American companies got the lion's share of the business. In recent years, however, the United States has become universally regarded as an unreliable supplier. France and Germany get nearly all of the business. The Soviet Union and Western Europe have become important international suppliers of fuel.

But perhaps the most disturbing effect of our national nonproliferation politics was that it caused us to lose most of our influence in international nonproliferation efforts. Before 1977, the United States played leading roles in all international programs and planning to discourage proliferation. We were the leading force in drawing up the international nonproliferation treaty and in getting it ratified by most nations of the world. We led the way in seeking and developing technological methods of assuring compliance, and of limiting problems. However, since the United States "went its own way," we have lost much of our credibility and have had much diminished influence in international nonproliferation programs.

In trying to understand the failure of the Carter effort to stop the spread of reprocessing technology, it is important to consider how effective it might be in stopping weapons proliferation. One obvious limitation was that it was designed only to obstruct the back country road, doing little to obstruct the two main highways to proliferation. Most of the world outside the United States recognized that point and, hence, regarded the Carter initiative contemptuously.

If a nation decides to develop nuclear weapons, lack of reprocessing facilities would hardly stop it. Fourteen nations now have such facilities, and others would have little difficulty in developing them. A commercial reprocessing plant designed to operate efficiently and profitably with minimal environmental impact is a rather expensive and complex technological undertaking, but the same is not true for a plant intended for military use where the only concern is obtaining the product. Construction of such a plant requires no secret information and no unusual skills or experience. Details of reprocessing technology have been described fully in the open literature. It was estimated[13] in 1977 that a crude facility to produce material for a few bombs could be put together and operated by five people at a cost of $100,000. A plant capable of longer-term production of material for eight bombs per year could be built and operated by 15 people, half of them engineers and the other half technicians, at a cost of $2 million. Either of these plants, or anything in between, could probably be built and operated clandestinely.

Thus, stopping reprocessing of commercial power reactor fuel is hardly an effective way of preventing weapons proliferation, and it is not widely viewed as such outside of the United States. On the other hand, reprocessing provides an important source of fuel for present reactors that could tide a needy nation over for a few years in an emergency. It is, furthermore, the key to a future system of breeder reactors which is the only avenue open to many nations for achieving energy independence. Unlike this country, with its abundant supplies of coal, oil, gas, rich uranium ores, and shale oil potential, and the like, many nations are very poor in energy resources. These include not only heavily industrialized nations like France and Japan, but nations aspiring to that status like Brazil, Argentina, and Taiwan. It is not difficult to understand why these nations are unwilling to trust their very survival to the mercy of Arab sheiks or the whims of American presidents for the indefinite future. They desperately want some degree of energy independence, and reprocessing technology is the key to the only way they can foresee of ever achieving it.

Above and beyond the practical difficulties U.S. nonproliferation policy has encountered, we might ask how important is its goal. There never was any hope that it could prevent a major industrialized nation from developing a nuclear weapons arsenal—there are now five nations with such arsenals. It could only hope to prevent a less-developed country like Brazil from taking such a step. By signing the international nonproliferation treaty, Brazil has renounced any such intentions. A Brazilian reprocessing plant would be subject to very close scrutiny by IAEA inspectors to see that its plutonium is not diverted for use in weapons. Add to this the facts that the plutonium it produces is ill-suited for use in weapons, and that a separate, secret plant could be built and used to produce weapons grade plutonium, and it seems clear that stopping a Brazilian reprocessing plant will not be the action that prevents that nation from developing nuclear weapons.

But suppose it did allow Brazil to develop a small arsenal of nuclear weapons—what could it do with it? It could threaten its neighbors, but they could easily be guaranteed against attack by the large nuclear weapons powers; Japan, Germany, and Scandinavia, for example, do not feel threatened by Russian or Chinese nuclear weapons because they are covered by the U.S. umbrella. There are few places in the world where a small nation could use a nuclear bomb these days without paying a devastating price for its action.

If one attempts to develop scenarios that might lead to a major nuclear holocaust, fights over energy resources such as Middle East oil must be at or near the top of the list. Anything that can give all of the major nations secure energy sources must therefore be viewed as a major *deterrent* to nuclear war. Reprocessing of power reactor fuel can provide this energy security, and therefore has an important role in *averting* a nuclear holocaust. That positive role of reprocessing is, to most observers, more important than any negative role it might play in causing such a war through proliferation of nuclear weapons.

After all of this discussion of proliferation, it is important to recognize that the use of nuclear power in the United States has no connection to that issue. If we stopped our domestic use of nuclear power, this would not deter a Third World nation from obtaining nuclear weapons, or conversely, use of nuclear power in the United States in no way aids such a nation in obtaining them. The only possible problems occur in transfer of our technology to those countries.

One of the most disturbing aspects of the proliferation problem is the utter lack of information on it that has been made available to the American public. I doubt if more than 1% of the public has any kind of balanced understanding of the subject. Based on the little information provided to them, most people have a distinct impression that our use of nuclear power adds substantially to the risk of nuclear war.

This impression has been cemented by the tactic of anti-nuclear activists to tie nuclear weapons and nuclear power together in one package, purposely making no effort to distinguish between the two. Consider this from an Evans and Novak column after the November 1982 election: "Eight states and the District of Columbia voted for a nuclear freeze [on weapons], but the one crucial issue on any ballot—Maine's referendum on [shutting down] the Yankee Power Plant—the pronukes won." The terms "anti-nuke" and "pronuke" are often used interchangeably in referring to nuclear weapons and power plants for generating electricity.

A TOOL FOR TERRORISTS?

A rather separate issue linking nuclear power with nuclear weapons is the possibility that terrorists might steal plutonium to use for making a bomb. This issue was first brought to public attention in 1973 in a series of articles by John McPhee in the *New Yorker* magazine later published as a book.[14] He reported on interviews with Dr. Ted Taylor, a former government bomb designer. Taylor had been worried about this problem for some time and had tried to convince government authorities to tighten safeguards on plutonium, which were quite lax at that time, but he could not stir the bureaucracy. The McPhee articles provided an instant solution to the lax safeguards problem—over the next 2 years, they were dramatically tightened. They also made Ted Taylor an instant hero of the antinuclear movement and the terrorist bomb issue stayed in the limelight for several years.

Let's take a look at that issue. To begin, consider some of the obstacles faced by terrorists in obtaining and using a nuclear bomb.[15] Their first problem would be to steal at least 20 pounds of plutonium, either from some type of nuclear plant or from a truck transporting it. Any plant handling this material is surrounded by a high-security fence, backed up by a variety of electronic surveillance devices, and patrolled by armed guards allowing entry only by authorized personnel. The plutonium itself is kept in a closed-off area inside the plant, again protected by armed guards who allow entry only to people authorized to work with that material. These people must have a security clearance, which means that they are investigated by the FBI for loyalty, emotional stability, personal associations, and other factors that might suggest an affinity for terrorist activities. When they leave the area where plutonium is stored or used, they must pass through a portal equipped to detect the radiation emitted by plutonium. It will readily detect as little as 0.01 percent of the quantity needed to make a bomb, even if it were in a metal capsule swallowed by the would-be thief. Plants conduct frequent inventories designed to determine if any plutonium is missing. In some plants these inventories are carried on continuously under computer control so as to detect rapidly any unauthorized diversion. There are elaborate contingency plans for a wide variety of scenarios.[16]

When it is transported, plutonium is carried in an armored truck with an armed guard inside. It is followed by an unmarked escort vehicle carrying an armed guard. All guards are expert marksmen qualified periodically by the National Rifle Association.

The truck and the escort vehicles have radio telephones to call for help if attacked, and they report in regularly as they travel. There are elaborate plans for countermeasures in the event of a wide variety of problems.[17]

The only significant transport of plutonium in connection with nuclear power would be of ton-size fuel assemblies in which the plutonium is intimately mixed with large quantities of uranium from which it would have to be chemically separated before use in bombs. If terrorists are interested in stealing some plutonium, it would be much more favorable for them to steal it from some aspect of our military weapons program where it is frequently in physical sizes and chemical forms easier to steal and much easier to convert into a bomb. That also would give them *weapons*-grade plutonium, which is much more suitable for bomb making than the reactor-grade plutonium from the nuclear power industry. It would be even better for them to steal some high-purity U-235 (which is not used in nuclear power activities) from our military program, since that is very much easier to make into a bomb. Of course, their best option would be to steal an actual military bomb.

It should be recognized that all of this technology for safeguarding plutonium is now used only for material in the government weapons program. There is essentially no plutonium yet associated with nuclear power. One might wonder how it would be possible to maintain such elaborate security if all of our electricity were derived from breeder reactors fueled by plutonium. The answer is that the quantities of plutonium involved would not be very large. All of the plutonium in a breeder reactor would fit inside a household refrigerator,* and all of the plutonium existing at any one time in the United States would fit into a home living room. The great majority of it would be inside reactors or in spent fuel, where the intense radiation would preclude the possibility of a theft. As in the case of radioactive waste, the small quantities involved make very elaborate security measures practical.

There have been charges that all these security measures with armed guards would turn this country into a police state. However, the total number of people required to safeguard plutonium would be only a small fraction of

*A large breeder reactor contains about 3,000 kg of plutonium which, in the form used (plutonium oxide), has a volume of 9 cubic feet. If all our electricity were derived from these reactors, about 400 would be needed, containing 3,600 cubic feet of plutonium, the volume of a living room 20 ft. × 18 ft. × 10 ft. high.

the number now used for security checking in airports to prevent hijacking of airplanes. That force has hardly given our country a police state character. If terrorists do manage to steal some plutonium from nuclear power operations and evade the intensive police searches that are certain to follow their theft, their next problem is to fabricate it into a bomb. Ted Taylor's assessment[18] of that task is indicated by the following quote:

> Under conceivable circumstances, a few persons, possibly even one person working alone, who possess about 10 kilograms of plutonium and a substantial amount of chemical high explosive could, within several weeks, design and build a crude fission bomb.
>
> By a "crude fission bomb" we mean one that would have an excellent chance of exploding with the power of at least 100 tons of chemical high explosive. . . . The key persons or person would have to be reasonably inventive and adept at using laboratory equipment. . . . They or he would have to be able to understand some of the essential concepts and procedures that are described in widely distributed technical publications concerning nuclear explosives, nuclear reactor technology, and chemical explosions [and] would also have to be willing to take moderate risks of serious injury or death.

Taylor suggested some available publications that would be useful. I read them, but the principal message I derived was that designing a bomb would be even more difficult than I had previously believed it to be. Perhaps some obscure statements in those publications contain the key to solving the problem; they would be readily recognized by a professional government bomb designer, but they were not at all recognizable to me.

In order to better understand the difficulties a terrorist would face in fabricating a bomb, let us consider opinions from some other professional government bomb designers. The following statement was obtained from J. Carson Mark of Los Alamos National Laboratory[19]:

> I think that such a device could be designed and built by a group of something like six well-educated people, having competence in as many different fields.
>
> As a possible listing of these, one could consider: A chemist or chemical engineer; a nuclear or theoretical physicist; someone able to formulate and carry out complicated calculations, probably requiring the use of a digital computer, on neutronic and hydrodynamic problems; a person familiar with explosives; similarly for electronics; and a mechanically skilled individual.
>
> Among the above (possibly the chemist or the physicist) should be one able to attend to the practical problems of health physics.
>
> Clearly depending on the breadth of experience and competence of

the particular individuals involved, the fields of specialization, and even the number of persons, could be varied, so long as areas such as those indicated were covered.

Note that this assessment is more optimistic from the public security viewpoint than Ted Taylor's statement. An even more optimistic assessment was given by E. M. Kinderman of Stanford Research Institute[20]:

> Several people, five to 10 with a hundred thousand dollars or so could do the job if the people were both dedicated to their goal and determined to pursue it over two years or more.
>
> One or a few competent physicist-engineers could probably arrive at a tentative design in a year or so.
>
> A chemical engineer and a metallurgist could...construct the essential equipment, make essential tests, and alone or with some help, operate a plant to produce the product dictated by the bomb designer.
>
> Others would be needed for the design and construction of the miscellaneous parts. . . . It is likely that the team will produce something with a force equivalent to 50 to 5,000 tons of TNT*...[and it] will weigh less than one ton.

Aside from the three statements quoted, the only other information from bomb experts with which I am familiar was a government statement[21]:

> ...a dedicated individual could conceivably design a workable device. Building it, of course, is another question and is no easy task.
>
> However, we also recognize that it is conceivable that a group with knowledge and experience in explosives, physics, metallurgy, and with the requisite financial resources and nuclear materials could, over a period of time, perhaps even build a crude nuclear explosive.

The principal information given to the public via the media was based on the separate efforts of an unnamed MIT student, and John Phillips, a Princeton student. The MIT student was hired for a summer to see how far he could get in trying to make a bomb, and the story was told on a NOVA television program. He produced a design and fabrication procedure on paper which a Swedish expert judged to have "a small but real chance of exploding with a force of 100-1000 tons [of TNT]." It was not stated whether this referred to a bomb made from weapons-grade, or reactor-grade, plutonium. It was also not clear what is meant by a Swedish expert, since Sweden has never built a bomb and has always professed no interest in such an undertaking.

A U.S. government expert who examined the MIT student's fabrication

---

*TNT is the chemical explosive referred to in the statement by Taylor quoted above.

procedure told me that anyone trying to follow it would almost surely be blown up by the chemical explosives long before his bomb was completed. Obviously the student had no experience in handling high explosives. This is an example of the various problems a terrorist would face. There is nothing secret about handling high explosives, but few people are experienced in this type of work. The handling includes cutting and shaping it, and attaching things to it—I have 30 years of experience in experimental physics and have mastered numerous techniques, but I would never consider undertaking anything so dangerous.

The effort by the Princeton student, John Phillips, was much more widely publicized. He made extravagant claims that his bomb would explode with a force of more than 10,000 tons of TNT. He took on a publicity agent, wrote a book, appeared on many TV and radio shows, received very wide newspaper coverage, and even ran for Congress.

What he claimed to have produced was a design for a bomb in a term paper prepared for a physics course. I spoke to his professor in that course, who said that there was nothing in his paper that would ordinarily be called a design. There were only crude sketches without dimensions. There were no calculations to support his claim that his bomb would work. He had collected a lot of information that would be useful in designing a bomb, for which the professor gave him an A grade.

Phillips was being called by media people so frequently that he had to have a separate telephone installed in his dormitory for that purpose. His professor told me that he himself had been contacted by many newsmen, but they never printed what he told them—they only trumpeted that Phillips had designed a workable bomb.

Several people have told me that professional government bomb designers have said that a design for a bomb by some student would work. I know that this could not be true because it would be a very serious breach of security regulations for a person who was ever involved with the government program to comment on a design that is available to the public. Note that the MIT student's design was judged by a "Swedish expert." With regard to such claims about Phillips' design, no professional could possibly consider a sketch without dimensions to be a design capable of being evaluated for performance. Science and technology are highly quantitative disciplines, but apparently Phillips does not understand that fact.

There have been numerous statements in newspapers, including our university paper, that any college student could design a nuclear bomb. In reply, I published an offer in our university paper of an unqualified A grade in both of the two courses on nuclear energy that I was teaching for any student who can show me a sketch of a workable plutonium bomb together with a quantitative calculation showing that it would work. My offer has been repeated about 10 times over the last 15 years. Three students turned in papers, but none of them had as much as 5% of what could be called a design.

All of this discussion has been about designs on paper, but as is clear from the above quoted statements by experts, that is only a small part of the task faced by terrorists. The fabrication requires a wide degree of expertise and experience in technical areas. It requires people capable of carrying out complex physics and engineering computations, handling hazardous materials, arranging electronically for a hundred or so triggers to fire simultaneously within much less than a millionth of a second, accurately shaping explosive charges, attaching them precisely and connecting the triggers to them, and so on. Where would terrorists find this expertise?

Experienced and talented scientists and technicians generally enjoy well-paid and comfortable positions in our society and hence are not likely to be inclined toward antisocial activity. Recruiting would have to be done under strictest secrecy, which would have to be maintained over the development period of many weeks. Even one unsuccessful recruiting attempt could blow their operation. Moreover, a participant would face a high risk of being killed in this work. And if the plot were discovered he would face imprisonment, not to mention an end to a promising career. Terrorists would surely face severe difficulty in obtaining the needed expertise.

But suppose, somehow terrorists succeeded in stealing the plutonium and making the bomb. Let us say it has the explosive force of 300 tons of TNT, which is an average of the various estimates by experts. What could they do with it?

We usually think of a nuclear bomb as something capable of destroying a whole city, but that refers to bombs many thousands of times larger. A bomb of this size would roughly be capable of destroying one city block, or one very large building. Ted Taylor uses,[18] as an example, the World Trade Center in New York City, which sometimes contains 50,000 people. That is the origin of the oft-quoted statement that one of these bombs could kill 50,000 people. It could also kill a similar number in a sports stadium by showering them with radiation and burning them with searing heat.

However, if killing 50,000 people is the terrorists' desire, there are many easier alternatives for accomplishing it. They could:

• Release a poison gas into the ventilation system of a large building.
• Dynamite the structural supports in a sports stadium so as to drop the

upper tier down on top of the lower tier; this should kill nearly all the people in both tiers.

- Discharge a large load of napalm (or perhaps even gasoline) on the spectators in a sports stadium, either by airplane or truck.
- Blast open a large dam; there are situations where this could kill over 200,000 people.
- Poison a city water or produce supply.

Any imaginative person could add many more items to this list.

Terrorists with a nuclear bomb would probably first try to use it as an instrument for blackmail. But nonnuclear tactics would be equally useful for that purpose. Unlike the nuclear bomb situation, there are plenty of people with all the know-how to carry out these actions.

Also unlike the nuclear bomb situation, we are doing nothing to avert their implementation, although there are many things we might do. We could guard ventilation systems of large buildings, but we don't. We could guard dams and reservoirs, but we don't. We could inspect sports stadium structural supports for dynamite before major events, but we don't. The high school I attended was recently rebuilt without windows, making its 3,000 students defenseless against poison gases introduced into its ventilation system. Terrorists could easily turn theaters or arenas into blazing infernos with blocked exits, but we don't guard against that. They could kidnap wives and children of Congressmen and other high officials, which might be very effective for their purposes, but we show no concern about that problem.

We are vulnerable to mass murder or blackmail by terrorists on dozens of fronts. All of them would be equally effective and infinitely easier for terrorists to take advantage of than making a bomb from plutonium stolen from the nuclear power industry. Yet there has been a great deal of concern expressed about the latter problem, and none about the others. Actually this may have served a useful purpose. Experts on terrorism have told me that it would be very favorable if terrorists devoted their energies to nuclear terrorism and were thus diverted from the easier and more destructive options available to them.

PLUTONIUM TOXICITY

Another property of plutonium unrelated to its use in bombs has attracted a great deal of attention. That is, its toxicity, as exemplified by Ralph Nader's statement that a pound of plutonium could kill 8 billion people.[22] Let's look into that question.

In Chapter 11 we showed how to calculate the toxicity of plutonium ingested into the stomach, which is the way it would most probably enter the human body if it is buried deep underground as part of radioactive waste. However, the most important health effects due to plutonium released from nuclear facilities occur when it becomes suspended in the air as a fine dust and is thereby inhaled into the lungs.*

It is straightforward to quantify the risks associated with this problem. When plutonium oxide, the form in which plutonium would be used in the nuclear industry and also its most toxic form, is inhaled as a fine dust, 25% of it deposits in the lung, 38% deposits in the upper respiratory tract, and the remainder is exhaled.[23] Within a few hours, all of that deposited in the upper respiratory tract, but only 40% of that deposited in the lung, is cleared out. The other 60% of the latter—(.25 × .60 =) 15% of the total inhaled—remains in the lung for a rather long time, an average of 2 years.

From the quantity of plutonium in the lung and the length of time it stays there, it is straightforward to calculate the radiation exposure to the lung in millirems. For example, a trillionth of a pound gives a dose of 1,300 mrem over the 2-year period (see Chapter 13 Appendix). From studies of the Japanese atomic bomb survivors, of miners exposed to radon gas, and other such human exposure experiences, estimates have been developed for the cancer risk per millirem of radiation exposure to the lung (see Chapter 13 Appendix).[24,25] Multiplying this by the sum of the radiation doses in millirems received by all those exposed gives the number of cancers expected. The result is that we may eventually expect about 2 million cancers for each pound of plutonium inhaled by people.[26,27] (By a more complex process, inhaled plutonium can also cause liver cancer and, to a lesser extent, bone cancer. Our treatment here is thus oversimplified.)

There is no direct evidence for plutonium-induced cancer in humans, but there have been a number of experiments on dogs, rabbits, rats, and

---

*If this seems contradictory to the discussion in Chapters 11 and 12, the following may be useful:

- For material suspended in air as a fine dust, the probability for it to be inhaled by a human is about 1 chance in a million.
- For plutonium buried deep underground, the probability that it will be inhaled by a human is about 1 chance in a trillion, a million times less.
- For plutonium buried deep underground, the chance that it will enter a human stomach is about 1 chance in 30 million.
- A quantity of plutonium inhaled is about 5,000 times more likely to cause cancer than the same quantity taken into the stomach, because in the latter case 99.99% of it is rapidly excreted. For most other materials, the difference between inhalation and dietary intake is very much less.

mice. The results of these are summarized[28] in Fig. 1, where the curve shows the expectation from our calculation. It is evident that the animal data give strong confirmation for the validity of the calculation.

The 2 million fatalities per pound inhaled leaves plutonium dust far from "the most toxic substance known to man." Biological agents, like botulism toxin or anthrax spores[29] are many hundreds or thousands of times more toxic. Plutonium toxicity is similar to that of nerve gas,[29] but given the choice of being in a room with equal quantities of plutonium dust and nerve gas, the latter would be infinitely more dangerous. It rapidly permeates the

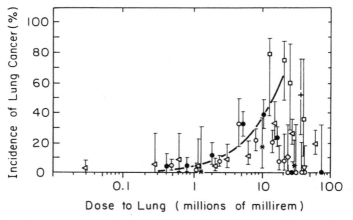

Fig. 1. Results of animal experiments with inhaled plutonium. The curved line shows the predictions of the calculation outlined in the text. Data are compiled in ref. 28, and the calculation is explained in detail in ref. 26.

□ PuO_2, dogs, J. F. Park, private communication to R.C. Thompson quoted in BNWL-SA-4911.

▽ PuO_2, mice, L. A. Temple *et al., Nature 183,* 498 (1959).

△ PuO_2, mice, L. A. Temple *et al., Nature 183,* 498 (1959).

◇ PuO_2, mice, R. W. Wager *et al.,* Hanford Report HW-41500 (1956).

○ Pu, citrate, rats, L. A. Buldakov and E. R. Lyubchanskii, translation in ANL-tr-864 (1970), p. 381.

● Pu pentacarbonate, rats, L. A. Buldakov and E. R. Lyubchanskii, translation in ANL-tr-864 (1970), p. 381.

◁ Pu nitrate, rats, R. A. Erokhim *et al.,* translation in AEC-tr-7387 (1971), p. 344.

+ Pu nitrate, rabbits, N. A. Koshurnikova, translation in AEC-tr-7387 (1971), p. 334.

* Pu pentacarbonate, rabbits, N. A. Koshurnikova, translation in AEC-tr-7387 (1971), p. 334.

room air, whereas plutonium, being a solid material, would be largely immobile.

In fact, it is rather difficult to disperse plutonium in air as a respirable dust. Individual particles tend to agglomerate into lumps of too large a size to be inhaled. In the experiments on animals, substantial effort and ingenuity was required to overcome this problem[30] and arrange for the plutonium dust to be inhaled.

The calculational procedure used here to obtain our result, 2 million deaths per pound inhaled, follows the recommendations of the International Commission on Radiological Protection (ICRP). It would be impossible to obtain a very different result without sharply deviating from them; at least three independent investigations have used them to evaluate the toxicity of plutonium[26,27] and they have all obtained essentially the same result. These ICRP recommendations are used by all groups charged with setting health standards all over the world, such as the Environmental Protection Agency and the Occupational Safety and Health Administration in the United States. They are almost universally used in the scientific literature.

Nevertheless, there have been at least two challenges to these procedures. The first was based on the so-called hot-particle theory, according to which *particles* of plutonium do more damage than if the same amount of plutonium were uniformly distributed over the lung, because the few cells near the particles get much larger radiation doses in the former case. The conventional risk estimates are based on assuming that the cancer risk depends on the average dose to all of the cells in the lung, while the hot-particle theory assumes it depends on the dose to the most heavily exposed cells.

This hot-particle theory had been considered by scientists from time to time over the years, but the issue was brought to a head when the antinuclear activist organization, Natural Resources Defense Council, filed a legal petition asking that the maximum allowable exposures to plutonium be drastically reduced, in view of that theory.[31] In response, a number of scientific committees were set up to evaluate the evidence. There were separate committees from the National Academy of Sciences, the National Council for Radiation Protection and Measurements, the British Medical Research Council, and others. All of them independently concluded[32] that there is no merit to the hot-particle theory, and that, if anything, concentration of the plutonium into particles is less dangerous than spreading it uniformly over the lung. The scientific evidence is too complex to review here, but a few points are easily understood:

- Animal experiments with much "hotter" particles of plutonium gave fewer cancers than those with normal particles containing the same total amount of plutonium.

• Particles are constantly moving from place to place, so during their 2 year residence in the lung, all cells are more or less equally irradiated.

• There were about 25 workers from Los Alamos National Laboratory who inhaled a considerable amount of plutonium dust during the 1940s; according to the hot-particle theory, each of them has a 99.5% chance of being dead from lung cancer by now, but there has not been a single lung cancer among them.[33]

As a result of these studies, the scientific community has rejected the hot-particle theory, and standard-setting bodies have not changed their allowable exposures. However, anti-nuclear activists continue to use the theory to justify such widely quoted remarks as "a single particle of plutonium inhaled into the lung will cause cancer."

Shortly after the fuss over the hot-particle theory had cleared away, John Gofman, head of a San Francisco anti-nuclear activist organization, proposed[34] a new theory of why plutonium should be much more dangerous than estimated by standard procedures. His theory was based on the idea that the cilia (hairs) that clear foreign material from the bronchial regions are destroyed in cigarette smokers, allowing the plutonium to stay there for years rather than being cleared within hours. He ignored direct experiments[35] that showed unequivocally that dust is cleared from these regions just as rapidly in smokers as in nonsmokers; apparently, to make up the difference, smokers have more mucous flow and do more coughing. In fact, if smokers cleared dust from their bronchial passages as slowly as Gofman assumes, they would die of suffocation. Because Gofman made errors in his calculation (as in using a surface of the bronchi that was 17 times too small), his paper was negatively critiqued by a number of scientists.[36,26] It has never gained any acceptance in the scientific community and has been ignored by all committees of experts and standard-setting groups. I know of no scientist other than Gofman who uses it in his work.

When my paper on plutonium toxicity[26] was first published, including its estimate of 2 million cancers per pound of plutonium inhaled, Ralph Nader asked the Nuclear Regulatory Commission to evaluate it. Judging from the number of telephone calls I received asking about calculational details, they did a rather thorough job, and in the end they gave it a "clean bill of health." Nevertheless, Nader continued to state, in his speeches and writings, that a pound of plutonium could kill 8 billion people, 4,000 times my estimate. In fact, he accused me[37] of "trying to detoxify plutonium with a pen."

In response, I offered to inhale publicly many times as much plutonium as he said was lethal. At the same time, I made several other offers for inhaling or eating plutonium—including to inhale 1,000 particles of plutonium of any size that can be suspended in air, in response to "a single particle...will cause cancer," or to eat as much plutonium as any prominent nuclear critic will eat or drink caffeine. My offers were such as to give me a risk equivalent to that faced by an American soldier in World War II, according to my calculations of plutonium toxicity which followed all generally accepted procedures. These offers were made to all three major TV networks, requesting a few minutes to explain why I was doing it. I feel that I am engaged in a battle for my country's future, and hence should be willing to take as much risk as other soldiers.

None of the TV networks responded (except for a request by CBS for a copy of my paper), so nothing ever came of my offer. However, antinuclear activists have used it to make me seem irrational—they say I offered to eat a *pound* of plutonium, whereas it was actually 800 milligrams, 550 times less. Some people have told me that antinuclear activists get so much media attention because they offer drama and excitement. It seems to me that my offer would have provided these, so there goes another explanation for why the media have been so unbalanced on nuclear power issues.

One story in connection with my offer gives insight into why journalists have performed so poorly in informing the public about radiation hazards. A national correspondent for the Dallas *Times Herald* quoted me as saying "I offered to inhale a thousand times as much plutonium as [Ralph Nader] would eat caffeine." I wrote a letter complaining about this and a host of other errors in his piece, to which his editor replied in part: "[You wrote] 'I offered to inhale a thousand times as much plutonium as he claims would be lethal, and to eat as much plutonium as he would eat caffeine'...this seems to be faithful to what our [correspondent] reported."

It is 5,000 times more dangerous to inhale plutonium than to eat it, and eating plutonium is about equal in danger to eating the same quantity of caffeine. Thus, if I were to do what the writer said I offered to do, I would be taking $(1,000 \times 5,000 =)$ 5 million times greater risk than Nader would be taking in eating the caffeine—I would surely be dead. Actually I offered to eat (not inhale) the same amount (not 1,000 times as much) of plutonium as he would eat caffeine, giving us equal risks. My offer to inhale plutonium was a completely separate item, intended to point out the ridiculousness of his statements about the dangers of inhaling plutonium. How a national correspondent can interpret my quote as he did, and how an editor can then fail to understand the difference when it is pointed out to him, is beyond my comprehension. Nevertheless, it is people like them, rather than the scientists, who are educating the public about radiation. Note that this is not a question of qualitative versus quantitative; being in error by a factor of 5 million is hardly a matter of lack of precision.

In evaluating the hazards from plutonium toxicity, it gives little insight to say that we can expect 2 million cancers per pound of plutonium inhaled unless we specify how much plutonium would be inhaled in various scenarios. This, of course, depends on the type of release, the wind and other weather conditions, as well as the number of people in the vicinity. But let us say that one pound of plutonium oxide powder gets released in the most effective way in an average big city location under average weather conditions.[26] In the hour or so before the wind blows dust out of the densely populated areas, only about 1 part in 100,000 would be inhaled by people,[38] enough eventually to cause 19 cancers. If people know about the plutonium, as in a blackmail situation, they could breathe through a folded handkerchief or piece of clothing, which would reduce the eventual death toll from 19 to 3. Better yet, they could go inside buildings and shut off outside air intakes for this critical short time period.

Eventually all of the plutonium dust would settle down to the ground, but there would still be the possibility of its later being resuspended in air by wind or human activities. Its ability to be resuspended is reduced by rain, dew, and other natural processes, as a result of which the principal threat from this process diminishes rapidly over the first year and essentially disappears thereafter.[39] All in all, this resuspended plutonium dust eventually causes about seven deaths.

Within a few years the plutonium works its way downward into the ground, becoming a permanent part of the top layers of soil. As is well known, it remains radioactive for a very long time. How much harm it does over that period depends on its probability of getting suspended in air by plowing, construction, or natural processes, and then being inhaled by humans. Due to these processes, an average atom of a heavy metal in the top eight inches of soil has 13 chances in a billion of being inhaled by a human each year.[26] If this probability is applied to the plutonium, it will cause a total of only 0.2 additional deaths over the tens of thousands of years that it remains radioactive.   '

During this period, plutonium in the soil can also be picked up by plant roots, thereby getting into food. This process has been studied in many controlled experiments and in various contamination situations such as bomb test sites and waste disposal areas.[40] Its probability is highly variable with geography, but even under the most unfavorable conditions, this would lead to less than one additional fatality over the tens of thousands of years. In summary, a pound of plutonium dispersed in a large city in the most effective way would cause an average of 19 deaths due to inhaling from the dust cloud during the first hour or so, with 7 additional deaths due to resuspension during the first year, and perhaps 1 more death over the remaining tens of thousands of years it remains in the top layers of soil. This gives an ultimate total of 27 eventual fatalities per pound of plutonium dispersed.[26]

It has often been suggested that plutonium dispersion might be used as an instrument for terrorism. But this is hardly realistic because none of the fatalities would occur for at least 10 years,* and most would be delayed 20 to 40 years. It could not be used for blackmail because if the dispersal is recognized, protective action is easily taken—breathing through handkerchiefs, or going indoors. Terrorists would do much better with nerve gas, which can be made from readily available chemicals; it leaves dead bodies at the scene.

There have been fears expressed that we might contaminate the world with plutonium. However, a simple calculation shows[26] that even if all the world's electric power were generated by plutonium-fueled reactors, and all of the plutonium ended up in the top layers of soil, it would not nearly double the radioactivity already there from natural sources, adding only a tiny fraction of 1% to the health hazard from that radioactivity. As is evident from the previous discussion, plutonium in the ground is not very dangerous, because there is no efficient mechanism for transforming it into airborne dust.

John Gofman, the anti-nuclear activist whose work has been discussed previously, has been speaking and writing about effects of plutonium toxicity on the basis of what he calls 99.99% containment.[41] By this he means that 0.01% of all plutonium used each year will be released as a respirable dust that will remain suspended in air for long time periods. It turns out that even the steel and asphalt industries do better than that in containing their products,[42] often holding respirable dust releases down to 0.001%. But in plutonium handling, releases are very much smaller. Let's consider the reasons for this.

In the steel and asphalt industries, the materials are heated far above the melting point, resulting in vigorous boiling, which is a prime mechanism for

---

*Lung cancers resulting from radiation exposure take *at least* 10 years to develop, and typically take about 30 years. This is the latent period discussed in Chapter 5.

converting some of the material into airborne dust; in the nuclear power industry, plutonium would not normally be heated to its melting temperature. During processing, steel and asphalt are handled in open containers, well exposed to the building atmosphere, whereas plutonium is always tightly enclosed and completely isolated from the building atmosphere. The air from the building atmosphere can mix with outside air only after passing through filters. In steel and asphalt plants these filters consist of ordinary fabric, whereas in plutonium plants they are specially developed high-efficiency filters capable of removing 99.9999% of the dust from air passing through them.

Current Environmental Protection Agency (EPA) regulations require that no more than about one part in a billion of the plutonium handled by a plant escape as airborne dust.[43] All plants now operate in compliance with that regulation. It is 100,000 times less than the releases Gofman has been assuming. If all of the electricity now used in the United States were derived from breeder reactors, the maximum allowable releases would be 0.0007 pounds per year. If all plants were in large cities, where we have shown that plutonium releases cause 27 deaths per pound, this would correspond to one fatality every 50 years somewhere in the United States. Since these facilities would not be in cities, the consequences would be considerably lower, much less than one death per century.

Of course, the EPA regulations do not cover releases in accidents, and there have been some of these. Two of the most notable were fires in a Rocky Flats, Colorado, plant for fabrication of parts for bombs, where plutonium is handled in a flammable form (the forms used in the nuclear power industry are not flammable).[44] In the earlier fire in 1957, about 1 part in 300,000 — .002 ounces — of the plutonium that burned escaped as dust.* After that, many improvements were made, so that in the much larger fire in 1969, only 1 part in 30 million of the plutonium that burned escaped. Safety analyses indicate that new improvements will considerably reduce even this low figure.

Since only a tiny fraction of all plutonium handled each year would be involved in fires or other accidents, much less than one-billionth of the total would be released. Thus, accidents are a much lesser source of plutonium in the environment than the routine releases which are covered by the EPA

---

*If 1 part in 100,000 of that was inhaled by humans, the total quantity inhaled was only about (.002/16 lb ÷ 100,000 =) one-billionth of a pound. With 2 million cancers per pound inhaled, the expected number of cancers is 0.002 — i.e., there is 1 chance in 500 for a single cancer to result. This was calculated for a plant in a city; the Rocky Flats plant is not in a city, so the actual number would be about 10 times smaller.

regulations, and the total impact of plutonium toxicity in a full breeder reactor electricity system would be less than one death per century in the United States. (This does not include releases in accidents, which are considered separately in Chapter 6. Plutonium plays a very minor role in them.)

The most important effects of plutonium toxicity by far are those due to nuclear bombs exploded in the atmosphere. Only about 20% of the plutonium in a bomb is consumed, while the rest is vaporized and floats around in the Earth's atmosphere as a fine dust. Over 10,000 pounds of plutonium has been released in that fashion by bomb tests to date,[44] enough to cause about 4,000 deaths worldwide. Note that the quantity already dispersed by bomb tests is more than 10 million times larger than the annual releases allowed by EPA regulations from an all breeder reactor electric power industry.

I am often asked why such tight regulations are imposed on plutonium releases if they involve so little danger. The answer is that government regulators are driven much less by actual dangers than by public concern. They do pay attention to technological practicalities, and it turns out not to be too difficult to achieve very low releases. Costs are taken into account, but all plutonium handling is now in the military program, where cost is not such an important factor. The guiding rule for regulators is that all exposure to radiation should be kept "As Low As Reasonably Achievable" (ALARA), and in the case of plutonium releases the regulation cited corresponds to EPA's judgement of what is ALARA.

The difficulty with this system is that the public interprets very elaborate safety measures as indicators of great potential danger. This increases public concern and perpetuates what has become a vicious cycle involving all aspects of radiation protection—the more we protect, the greater the public concern; and the greater the public concern, the more we must protect.

One often hears that in large-scale use of plutonium we will be creating unprecedented quantities of poisonous material. Since plutonium is dangerous principally if inhaled, it should be compared with other materials which are dangerous to inhale. If all of our electricity were derived from breeder reactors, we would produce enough plutonium each year to kill a half trillion people.* But as has been noted previously in Chapter 5, every year we now produce enough chlorine gas to kill 400 trillion people, enough phosgene to kill 18 trillion, and enough ammonia and hydrogen cyanide to

---

*Each reactor produces about 500 kg of plutonium per year, so the 400 reactors needed would contain $(400 \times 500 =) 2 \times 10^5$ kg. Since plutonium could kill 2.3×10^6 people per pound[27] if all of it were inhaled by people, the potential toll is $(2 \times 10^5 \times 2 \times 10^6 \cong) 4 \times 10^{11}$, a half trillion.

kill 6 trillion with each. It should be noted that these materials are gases that disperse naturally into the air if released, whereas plutonium is a solid that is quite difficult to disperse even intentionally. Of course, plutonium released into the environment will last far longer than these gases, but recall that the majority of the harm done by plutonium dispersal into the environment is due to inhalation within the first hour or so after it is released. The long-lasting nature of plutonium, therefore, is not an important factor in the comparisons under discussion.

One final point about plutonium toxicity that should be kept in mind is that all its effects on human health that we have been discussing are *theoretical*. There is no direct evidence or epidemiological evidence, that the toxicity of plutonium has ever caused a human death anywhere in the world.

I have been closely associated professionally with questions of plutonium toxicity for several years, and the one thing that mystifies me is why the antinuclear movement has devoted so much energy to trying to convince the public that it is an important public health hazard. Those with scientific background among them must realize that it is a phony issue. There is nothing in the scientific literature to support their claims. There is nothing scientifically special about plutonium that would make it more toxic than many other radioactive elements. Its long half life makes it *less* dangerous rather than more dangerous, as is often implied; each radioactive atom can shoot off only one salvo of radiation, so, for example, if half of them do so within 25 years, as for a material with a 25-year half life, there is a thousand times more radiation per minute than emissions spread over 25,000 years, as in the case of plutonium.

No other element has had its behavior so carefully studied, with innumerable animal and plant experiments, copious chemical research, careful observation of exposed humans, environmental monitoring of fallout from bomb tests, and so on. Lack of information can therefore hardly be an issue. I can only conclude that the campaign to frighten the public about plutonium toxicity must be *political* to the core. Considering the fact that plutonium toxicity is a *strictly scientific* question, this is a most reprehensible situation.

I am convinced that the public has bought the propaganda about the dangers from plutonium toxicity. Ask a layperson; he or she will probably tell you that plutonium is one of the most toxic substances known and is a terrible threat to our health if it becomes widely used. The media accept this as fact; plutonium toxicity is no longer treated as an issue worthy of their attention. The deception of the American people on this matter is essentially complete. Lincoln was wrong when he said, "You can't fool all the people all the time."

Chapter 14 / THE SOLAR DREAM

One final point of public misunderstanding is the widespread impression that solar electricity will soon be replacing nuclear power, so there is no need to bother with nuclear energy. There are vociferous political organizations pushing this viewpoint, and a substantial portion of the public seems to be largely convinced that our primary source of electricity in the next century will be solar.

As a frequent participant in meetings on energy technology, I have come to know several solar energy experts, but I have yet to meet one who shares the above opinion. Their professional lives are devoted to development of solar electricity, and most of them are very enthusiastic about its future. Nevertheless, they have encouraged me in my efforts on behalf of nuclear power, saying frankly that the public's expectations for solar power are unrealistic. They foresee its future, at least for the near term, as a *supplement* to other technologies, with advantages in certain situations, rather than as the principal power source for an industrialized society.

My purpose here is to explain why that is so. It should be understood from the outset that I am not in any sense an expert on solar electricity. I have never done research in that field or even in the basic physics of semiconductors on which much of it is based. My knowledge is derived from relatively shallow reading, brief conversations with scientists working in the field, and preparing lectures for classes that I teach.

The materials that I read, mostly obtained from the U. S. Department of Energy and Solar Energy Research Institute, and the people from those organizations that I talk to, are always highly enthusiastic. They describe new records in efficiency of solar cells, ideas for new materials that have potential for much improved performance, reduced costs, new records in production and sales, new applications being undertaken, and the like. This is as it should be for a new and rapidly developing technology. If it were not, the technology would never have a chance. It reminds me of the early days in development of nuclear power, when enthusiasts thought it would be so cheap that it would drive all other energy sources out of the market. This early enthusiasm is a wonderful thing, but it should not be confused with what can be counted on to happen. Some technologies, like electronics and computers, have fulfilled the expectations of their early enthusiastic promoters, but the much more usual situation is for them to find useful niches but to fall far short of these dreams.

A recent and somewhat relevant example is the use of solar energy for heating and cooling. In the mid-1970s, enthusiasts were predicting that solar collectors would soon replace furnaces and hot water heaters in every home. New businesses were springing up everywhere for manufacturing and marketing solar heating systems and improved models kept coming out. Rampant rumors circulated about radical new improvements in the offing. I took my classes on tours of solar-heated buildings, and brought in experts on solar heating to lecture in our university.

Since the technology of solar heating is relatively simple, it matured very rapidly, and by the early 1980s it became reasonably clear that development had gone about as far as it could go. Well-developed marketing organizations made every effort to sell their products, aided by unprecedented subsidies. The federal government subsidized 40% of the purchase cost of any solar equipment, and the states added additional subsidies up to 30%, leaving only 30% of the cost to be paid by the home owner. I actually had a pen in my hand ready to sign a contract for installation of a solar water heater in my home, but I decided to delay until I did a careful cost calculation. I found that the cost of heating water in my home was 3 times as high with the solar energy system as with natural gas. I became interested in the subject, did research, and eventually published a paper on it.[1] My conclusion was that in the northeastern United States the cost per million BTU (the BTU is the usual unit of heat energy) was about $8 from natural gas and $13 from heating oil versus $27 for solar water heating and $72 for solar space heating. But the best way to save money was to insulate water heaters, which costs only $5 per million BTU saved.

The solar-heating product was just too expensive, even with the government subsidies. The public did not buy it, and the industry is dying. The dreams of the early enthusiasts did not materialize.

That doesn't mean that solar energy cannot be important for generating electricity, and it may well become very important. Development in that area has been highly encouraging. There are two basic problems that must be overcome. One is that the cost is now much higher than for electricity generated by conventional sources, and the other is that no electricity can be generated when the sun isn't shining.

There are situations in which these problems are not important, and they have helped to keep the enthusiasm thriving. One example is pumping water for irrigation in an area not serviced by the power grid; it doesn't matter if the operation stops when the sun isn't shining. Another example is in underdeveloped nations where most of the country is not serviced by the power grid, leaving expensive diesel engine generation as the only alternative. The diesel fuel is often scarce and expensive to transport, so it can be saved for night-time use while utilizing solar electricity during the day. Even in this country, many buildings not easily connected to a regular power grid use solar electricity. These applications are already being exploited, and as solar electricity becomes cheaper, many more will become practical.

But the best hope for truly large-scale application is for utilities to use solar energy to service peak loads during hot summer afternoons when a great deal of electricity is needed for air conditioning. This is usually a time when the sun is shining brightly, making lots of solar energy available. Since utilities normally use internal combustion turbines to service these peak demands, solar electricity would not be competing directly against the lowest-cost conventional electricity sources.

Of course, solar electricity need not compete with conventional forms everywhere to be useful. It need only be competitive in areas where there is plentiful sunshine and a strong need for air conditioning, and where coal and gas are not available and oil is expensive because of difficult transport requirements. There may be situations where air pollution control regulations inhibit use of coal or oil, or where communities are not willing to depend on oil because of fears about supply reliability. With all of these applications in mind, solar electricity enthusiasts are optimistic about "penetrating" the utility market before the end of this century.

In our discussion we will consider only the three principal approaches to utilizing solar energy for generating electricity: (1) photovoltaics, popularly known as solar cells; (2) solar thermal facilities, in which sunlight from a large area is collected by mirrors that focus it onto a receiver which is thereby intensely heated, with this heat used to generate electricity; and (3) wind turbines, an elaboration of the familiar windmill (since winds are driven by forces generated by the sun's heat, wind power is often classified as solar energy). Many other solar technologies could be discussed here, such as ocean thermal gradient, satellite solar, tidal, ocean waves, and biomass. Each of these has had its own proponents, but most solar energy experts consider them to be of lesser importance, and relatively little emphasis is now being put on developing them.

PHOTOVOLTAICS

There are several variations of photovoltaics involving different degrees of tracking the motion of the sun and using lenses to concentrate the sun's rays. Each has advantages, but at an added cost, which roughly compensates for them. We, therefore, simplify our discussion by considering only one of the options, a fixed flat plate. According to the directors of the government program, for photovoltaics to penetrate the utility market,[2] the module cost must be reduced to $45 per square meter if the efficiency is improved to 15%, or to $80 per square meter if the efficiency can be raised to 20%, assuming that the system life expectancy can be extended to 30 years (all costs are in 1987 dollars). These are the program goals. In 1982, the best performance was an efficiency of 9.8%, a cost of $1,140 per square meter, and 15 years life expectancy. In 1987, this had improved to 12% efficiency, a cost of $480 per square meter, and a life expectancy of 20 years.[3] If these trends continue—i.e., if every 5 years the efficiency improves by 2%, the cost is cut in half, and the life expectancy is increased by 5 years— photovoltaics will penetrate the utility market by the end of the century.[4] Those involved in the development are very confident that this will happen.

Let's try to understand what a marvelous accomplishment this would be. An ordinary cement sidewalk costs about $30 per square meter. The program goals are, therefore, to be able to cover the ground with solar cells for about twice the cost of covering it with a layer of cement. A solar cell is a highly sophisticated electronic device, about one inch in diameter, based on advanced principles of quantum physics developed in the 1950s. It is made from materials of extremely high purity, a purity that was unattainable even in scientific laboratories until the late 1940s. Cells are manufactured by processes that have taken some of the best efforts of modern technology to develop. They must be capable of standing up to all the vagaries of outside weather for 30 years. To cover the ground with these sophisticated devices for just twice the price we pay to cover it with a thin layer of cement, manufactured simply by grinding up rock and heating it, would indeed be an impressive accomplishment.

The numbers quoted here also explain why solar electricity is so expensive—the energy in sunlight is spread so diffusely that we must collect it from large areas with correspondingly large collectors in order to obtain appreciable amounts of power. To produce the power generated by a large nuclear plant would require covering an area 5 miles in diameter with solar cells.

There is a limit to the miracles that can be achieved in solar cell development. In addition to the cost goals quoted above for photovoltaic cells, there is a somewhat larger cost for mounting, electrical hook up, power conditioning, and other standard engineering operations that cannot be easily reduced. It therefore seems unlikely that an operating solar power plant can ever cost less than $1,000 per peak kilowatt. Since their power output averaged over day and night is only about 20% of the peak, this corresponds to a cost of $5,000 per average kilowatt. The cost estimate for the new generation of nuclear power plants is under $2,000 per average kilowatt. A nuclear plant requires fuel and more operating and maintenance labor, but two-thirds of the price of their electricity is in the original plant construction cost.

SOLAR THERMAL ELECTRICITY

Solar thermal facilities are of various varieties, some of them designed for small applications and others designed to produce very high temperatures for special applications. Here we will consider only *central receivers,* in which the sunlight falling on many acres of land is focused by mirrors onto a single receiver. The fluid in the receiver is thereby heated, and this heat is used to produce steam that drives a turbine to generate electricity, as in a nuclear or coal-burning power plant. The mirrors must be moved under computer control to very accurately track the motion of the sun across the sky in a path that changes day by day. Photovoltaics can operate on diffuse light, producing an electrical output proportional to the total amount of light striking them, but that is not the case here. An image of the sun coming from each mirror must fall accurately on the central receiver, which means that when a cloud covers the sun, no energy is collected. Clearly, this is a facility only for very special locations where clouds are unusual.

A system generating 10,000 kW of electricity, including over 1,800 mirrors, each roughly 20 feet on a side, has been constructed and successfully operated since 1984 at Barstow in the California desert.[5] Several smaller facilities are also in operation, largely for research purposes. Improved methods have been developed for transferring the heat from the receiver, at the top of a 300-foot-high tower in the Barstow plant, to the steam generation facility on the ground. There has been great progress in mirror development, increasing the area of individual glass-metal mirrors while reducing their costs, and introducing stretched-membrane mirrors, which have a potential for further cost reductions. With current technology, a plant could be constructed for $3,000 per peak kilowatt versus the $11,000 per peak kilowatt cost for the Barstow plant. Government program directors estimate that the utility market will be penetrated if the cost gets down to $1,000 per peak kilowatt. Enthusiasts believe that this can be achieved by 1995.

If all goes well, solar thermal plants may be contributing to service of peak power loads in the southwest desert by the turn of the century. But their potential is limited. It is almost impossible to imagine them being built in the northeast or midwest, where cloudiness is common. Photovoltaics are based on the quantum physics of materials, which is a field where "miracles" have occurred before and are always a possibility for the future. But solar thermal technologies are relatively standard engineering developments, which is hardly an area where one can hope for "miracles," especially after well over a decade of effort. Solar enthusiasts are, therefore, much more hopeful for the future of photovoltaics, and that is where most of the research effort is now being directed.

WIND TURBINES

Windmills have been used for many centuries to grind grain and pump water, and they were widely used to generate electricity on farms early in this century. But with the extension of power lines into rural areas in the 1930s, the enterprise collapsed. As a result of the energy crisis of 1974 and the big increase in energy costs that followed, new initiatives were undertaken under government sponsorship.[6] Initially, there were unexpected problems with structural weakness, vibrations, and noise, and efficiencies were not as high as expected. A sizable research and development program was, therefore, undertaken to improve understanding. In order to encourage use of wind, a law was passed requiring utilities to pay for any power fed into their lines by owners of wind turbines. Substantial tax credits were offered to equipment purchasers.

About 92% of all U.S. energy now generated by wind is in California, where, by the end of 1987, there were 16,000 wind turbines in operation with a total capacity of 1,440,000 kW, more than the capacity of a large nuclear plant. However, the electricity produced was only 25% of what would be produced by that nuclear plant, because the average wind speed is far below the maximum wind speed on which the capacity of the turbines is based.

By the mid-1980s, lifetime reliability issues emerged as a primary concern, and fatigue of materials became a vital research problem which is now becoming better understood. The present assessment is that in order for this technology to become broadly competitive with fossil fuels (1) reliability and durability must be improved to achieve an expected lifetime of 30 years, (2) costs must be reduced by 25%, and (3) efficiencies must be increased by about 50%. Enthusiasts believe these goals can be achieved by the mid-1990s.

It would take 50,000 wind turbines of the type being used in California to replace the average electricity produced by a single nuclear or coal-burning plant, which is hardly an inviting prospect for an electrical utility. Management costs would be horrendous. Important efforts are, therefore, being made to develop larger systems. The largest to date is a 3,200 kW, 320-foot in diameter turbine at Kahuku, Hawaii. Conditions in Hawaii are especially well suited for use of wind; moreover, coal or oil must be imported from long distances, making them expensive. A blade the length of a football field mounted on a tower and turning in the wind is a relatively major installation, but it would still take a thousand of these to produce the average electricity generated by a single nuclear or coal-burning plant.

An important disadvantage of wind turbines is that they are not well adapted to servicing peak loads on hot summer afternoons, since winds blow mostly on winter nights. A wind turbine should, therefore, be thought of only as a fuel-saving device. It does not reduce the need for other power plants. For remote areas not serviced by utility power lines, however, wind turbines are already quite useful. They are well suited to pumping water for irrigation.

Persistent high wind speeds are, of course, an essential requirement, and these are not to be found in many areas. However, as efficiencies of wind turbines are improved, more locations will become suitable.

Intermittent Availability

Since electrical energy is difficult to store, it must ordinarily be used as it is produced. Our principal uses, unfortunately, do not vary in time in the same way as sunshine and wind vary. There is little seasonal variation in our use of electricity, but the influx of solar energy is 2 to 3 times higher in summer than in winter. Therefore, if all our electricity were solar, the capacity would have to be large enough to provide the 24-hour needs even during short, dull winter days, leaving most of it idle and unproductive during the long, sunny days of summer. A more difficult problem is that we use a great deal of electricity at night when there is no sunshine. The most obvious solution is *storage*, and in this connection we first think of storage batteries like those used in automobiles. These could solve some of the short-term problems; for example, $6,000 worth of batteries replaced every 2 years could be charged during the day enough to handle ordinary nighttime uses in a home without air conditioning. The problem would be much more difficult for businesses and factories that use a lot of electric power at night and in winter. These costs would clearly be unacceptable.

The only other practical storage system for electricity is using it to pump water up to a reservoir on a hill; it can then generate electricity when it is allowed to flow back down. This is expensive to construct, wastes about one-third of the electricity, has various environmental problems associated with flooding large land areas, and is applicable only in places with plenty of water and hills. The most important project of this type ever undertaken, the Storm King reservoir on the Hudson River north of New York City, was delayed for many years and finally abandoned because of opposition from environmental organizations.

Aside from storing electricity, the other simple solution to variations in availability is to have back-up sources of electric power. One might consider having a nuclear or coal-burning power plant available, but this would make no sense economically. The standby plant would have to be constructed, and it would need nearly a full complement of operating and maintenance personnel. The only saving by using solar energy would be in fuel costs, which represent only about 20% and 50%, respectively, of the total cost of nuclear and coal-fired power. The cost of back-up power to the customers would have to be not much less than the cost of obtaining *all* their electricity from those plants.

We often hear stories about individuals with windmills or solar cells using a regular utility line for back-up power and selling the excess power they generate at various times back to the utility—utilities are required by law to purchase it. This does little harm as long as only a few individuals are involved, but it wouldn't work if a large fraction of customers did it. The utility would not only have to build and maintain back-up power plants without selling much of their product, but they would have to buy a lot of power they don't need when the sun is shining or the wind is blowing. The utility could survive only by raising the price of the electricity it does sell sky high.

In general, the amount of back-up or storage capacity needed to overcome the variable nature of sunshine and wind depends on how much inconvenience we are willing to endure. But even to approach the dependable electrical service we now enjoy would be extremely expensive.

In order to improve on this situation, the U. S. Department of Energy is supporting a substantial research effort to develop improved batteries.[7] The most promising one is the sodium-sulfur battery, which operates at high temperatures—about 600°F. Like all batteries, it consists of many individual cells connected together, and if a few of these fail, the efficiency drops dramatically. At early stages of the development, 60% of the cells were failing after 250 chargings, but this has now been reduced to 5%. The goal of the development program is to achieve 80% efficiency, 2,500 charge-discharge cycles, and a cost of about $90 per kilowatt-hour. For one charge-discharge cycle, the cost would then be ($90/2500 × 0.8 =) 4.5 cents per kilowatt-hour, roughly equal to the costs of electricity generated by coal or nuclear power given in Chapter 10. If this goal is achieved, we still must add the cost of installation and maintenance. This would make the cost of storage alone somewhat higher than the present total cost of electricity. Clearly something much better is needed.

SOLAR VERSUS NUCLEAR POWER

The electric power requirements that a utility must supply consist of the sum of (1) *base load*, which continues day and night and accounts for about two-thirds of all electricity used; (2) *intermediate load*, which is the increase above the base load that is normally encountered during most of the day and early evenings; and (3) *peak load*, which occurs for a few hours on most days and for longer times in special circumstances, like on very hot days when there is abnormally heavy use of air conditioning, or on exceptionally cold days in areas where electric heating is in widespread use.

Nearly all of the cost of nuclear power is in the construction of the plant. Fuel costs are much lower than for fossil fuel plants. It therefore pays to operate nuclear plants whenever they are available; consequently, they are used to provide base load service. A utility would not build more nuclear plants than it needs for its base load because it cannot afford to have a nuclear plant sit idle. For electricity derived from fossil fuel steam plants, 50-75% of the cost is due to the cost of the fuel. Hence, there is a substantial savings in shutting them down at night when the power in not needed. This makes them well suited for servicing intermediate load. Where nuclear plants are not available, they are also used for base load.

Since fossil fuel steam plants are still rather expensive to construct, it does not pay to build them for peak load service; they would sit idle the great majority of the time. Peak load service is normally provided by internal combustion turbines, which are relatively inexpensive to purchase but are inefficient, use expensive fuel, and hence are costly to operate.

Solar electricity is most ideally suited to providing peak load power due to heavy use of air conditioners on hot, sunny days during the summer. In that peak load application, solar will compete mostly with the internal combustion turbine, which is the most expensive source of electricity now in use. The optimism of solar electricity enthusiasts for penetrating the utility market by the end of the century is based on the prospect of succeeding against that competition. Since using solar energy to replace internal combustion turbines would reduce our use of oil and of machinery that is largely imported, it is very much a socially desirable goal.

Since solar energy is available during the daytime, it also has the potential of competing for service of the intermediate load. Of course, solar plants installed to supply peak power would be used all year and would contribute to the intermediate load. Since they consume no fuel, it saves money to use them—but this contribution would be relatively minor. However, since steam plants normally used to service intermediate loads are much more efficient and use cheaper fuel than internal combustion turbines, solar competition for the bulk of intermediate load will be much tougher. Moreover, intermediate load is normally as high in winter as in summer, so it is wintertime solar energy, which is much less available, that must compete here. However, in areas where coal is not available, as in most of our coastal areas, where a substantial fraction of the population is located, this competition would become easier if oil prices rise sharply. This might allow photovoltaics to penetrate the utility market for the bulk of intermediate load service. Solar electricity may thus serve as a check on price increases imposed on us by OPEC, a highly desirable goal.

If environmental restrictions on coal burning should become really severe, the price of that technology might escalate. This would improve the competitive position of solar energy for the rest of the intermediate load. Since solar energy is much less harmful to the environment than fossil fuel burning, and since it would be highly desirable to save these fossil fuels for other uses (see Chapter 3), it would be socially desirable for solar energy to take over as much of the intermediate load as possible. This hope is included in the dreams of the solar electricity enthusiasts, and we must wish them well in these endeavors.

The situation with regard to base load service is very different, however. Since base load electricity is the lowest in cost, the price competition becomes much stiffer. But more importantly, solar electricity can only contribute here in combination with a very large capacity for electrical energy storage, presumably with batteries. Even if current development goals are achieved, this storage alone will be more expensive than current base load electricity. Hence, for solar electricity to compete for base load service, two independent "miracles" would be needed, one in drastically reducing the cost of solar electricity, and the other in substantially reducing the cost of batteries beyond present program goals. Of course, if electricity storage should become really cheap, nuclear power would be in a position to compete for peak and intermediate load service.

Since nuclear power is used only for base load service, there will be no competition between nuclear and solar electricity in the foreseeable future. Each has its place in our nation's energy mix, and these places are very different. Each serves to reduce the environmental problems from fossil fuels discussed in Chapter 3. Each serves to alleviate the political and economic problems incurred in importing oil. The fact that nuclear and solar electricity are not in competition and probably never will be is the bottom line. If I were to return to Earth thousands of years from now, long after fossil fuels are gone, I would not be surprised to see nuclear reactors generating base load power and photovoltaics providing the intermediate and peak loads.

ENVIRONMENTAL IMPACTS OF SOLAR ELECTRICITY

Even if there were a competition between solar and nuclear electricity, there is no technically valid reason to prefer the former. It was pointed out previously that production of the materials for deploying a solar cell array requires burning 3% as much coal as would be burned in generating the same amount of electricity in coal-burning power plants. Roughly the same is true for the power tower and wind turbine applications of solar energy. That means that they produce 3% as much air pollution as coal burning. This is not a great environmental problem, but it still makes them more harmful to health than nuclear power. In addition, there are long-term waste problems, discussed in Chapter 12, which pose many times more of a health problem than the widely publicized nuclear waste. There are lots of poisonous chemicals used in fabricating solar cells, such as hydrofluoric acid, boron trifluoride, arsenic, cadmium, tellurium, and selenium compounds, which can cause health problems. Also, there is much more construction work needed for solar installations than for nuclear; construction is one of the most dangerous industries from the standpoint of accidents to workers.

If photovoltaic panels on houses become widespread, how many people would be killed and injured in cleaning or replacing solar panels on roofs, or in clearing them of snow? What about the dangers in repairing the complex electric conversion systems? Over a thousand Americans now die each year from electrocution, and the power-conditioning equipment needed for a solar electricity installation would represent a major increase in this risk. Back-up systems, most especially diesel engines in the home, have serious health problems. Diesel exhausts include some of the most potent carcinogens known, and they contribute to most of the other air pollution problems discussed in connection with coal burning in Chapter 3.

Large solar plants also create environmental and ecological problems. What happens to the land and animals that live on it when a 5-mile diameter area is covered with solar cells or mirrors? Desert areas, which are most attractive for solar installations, are especially fragile in this regard.

Wind turbines are noisy, and some consider them to be ugly, especially if they dominate the landscape for many miles in every direction—recall that it takes a thousand very large installations to replace one nuclear plant. Central receivers use a great deal of water, which is generally in short supply in deserts where these installations would be most practical. All solar electricity technologies require a lot of land, inhibiting its use for other purposes.

SOLAR VERSUS NUCLEAR POLITICS

In spite of the facts cited in the last section, the public considers solar energy to be the safest, soundest, and most environmentally benign way of obtaining electricity. It also seems to believe that solar electricity will be very cheap and will eventually replace nuclear power for that reason alone.

Everything about solar energy is viewed as good and desirable. I often wonder why this is so, since sunshine is the principal cause of human cancer. Even if we consider only fatal cancers, it is worse than radiation in this regard.

This favorable public image of solar energy has been very important politically. The Carter Administration went to great lengths to promote it, while nuclear energy was officially labeled "the source of last resort." The Carter appointee as head of TVA (Tennessee Valley Authority), our nation's largest electrical utility, refers to the present time as "the presolar age." Government agencies stumbled over one another in rushing to distribute prosolar propaganda, while any public distribution of information favorable to nuclear power met with harsh criticism from Congressmen and administration officials. My use here of the word "propaganda" is intentional. I call the solar material propaganda because it created false impressions by not mentioning the very serious cost and intermittent availability problems discussed above and by implying that our electricity has an exclusively solar future.

This mixing of science and politics is a dangerous tendency more suited to a banana republic than to a nation so heavily dependent on technology. Unfortunately, nuclear science has not been immune to it. Liberal Democrats seem to be against nuclear power, while conservative Republicans seem to favor it. This split does not apply to the involved scientists, most of whom are politically liberal Democrats, who on the basis of their scientific knowledge, strongly favor nuclear power. I personally have been a liberal Democrat all of my life, as has every member of my family for over 60 years. I, as well as the majority of my scientific colleagues, are passionately devoted to the welfare of the common people (this is my favorite definition of a "liberal"). It is clear to us that their welfare is heavily dependent on a flourishing nuclear power program.

Liberal Democrat politicians have not always opposed nuclear power. These days, people view nuclear power as being favored by the utilities, with lukewarm support from government bureaucrats, against the resistance of an unwilling body politic and Congress. But in the past, it was quite the opposite. Congress, under the leadership of its House-Senate Joint Committee on Atomic Energy, was the original driving force behind development of nuclear power in the 1950s. The government bureaucracy represented by the Atomic Energy Commission, favored a slow, drawn-out development program, while the utility industry was heavily resistive. In fact, the utilities were brought into the program only with the threat that if they didn't cooperate, the government would develop its own nuclear power program to compete with their coal- and oil-fired plants. The early history of nuclear power development was punctuated by strong pressures from Congress to speed things up. During this period and up to the early 1970s, such well-known liberal Democrat Senators as John Pastori, Clinton Anderson, Henry Jackson, Albert Gore, and Stuart Symington served on the Joint Committee and played key roles in promoting nuclear power. When information critical of it appeared on the scene, the Joint Committee held hearings for which they called in prominent scientists to explain the facts. Intelligent deliberations were held among well-informed and mutually respecting people, and questions were thereby settled.

The opposition to nuclear power among politicians started in the early 1970s when they stopped taking advice from the scientific establishment and instead began taking it from political activist groups belonging to what is generally referred to as "the environmental movement." These groups, principally led by Ralph Nader, used or desired little scientific information. They largely formulated their beliefs on the basis of political philosophy, and then found scientists wherever they could to support them, without regard to the consensus of opinions throughout the scientific community.

Somehow, their political philosophy told them that nuclear energy is bad and solar energy is good. They built a case to support that position as lawyers often build cases—the only object was to win. Facts and scientific information were introduced or ignored in accordance with how they affected their case. They weren't seeking the truth—they were sure they knew it. They were only seeking material to help them convince the public.

They sold their case to many liberals. "Solar is good, nuclear is bad" became an article of faith in the liberal establishment, and the public was soon convinced by the barrage of propaganda that followed. In this process, the scientific experts were essentially ignored. We scientists are inexperienced and inept at political gamesmanship. Still we have fought, and will continue to fight, to get our message to the public. It has been a long, uphill battle.

Our case is based on science, while the opposition is based on political philosophy. When a nation whose welfare is highly dependent on technology makes vital technological decisions on the basis of political philosophy rather than on the basis of science, it is in mortal danger.

Before closing this section on politics, I should mention that there are some people who foresee the day when all our electricity will be solar, but they envision a very different world than our present one. It is a world of low technology and a simpler life, a more desirable lifestyle, in their view. It will of necessity be a life of more unmechanized farming and manual labor, of fewer machines, comforts, and conveniences. They call it living in harmony with nature, but it might also be called sliding back toward the lifestyle of our primeval ancestors. In such a world, they contend that there would be no place for large nuclear or coal-fired power plants, and little place for other large industrial operations except, presumably, for manufacturing solar cells.

The validity of their views depends on social, political, demographic, and psychological considerations on which I have no claim to expertise. I only want it to be clearly understood that their ideas are driven by political considerations rather than by scientific, technical, or economic analyses. Their attempts at the latter have been shallow and heavily biased and have generally received rebuttal rather than acceptance by experts.

CONCLUSIONS

It would be nice to conclude this chapter with a statement that politics doesn't matter, that scientific and economic facts make it clear that nuclear and solar electricity are not in competition, and that they probably never will be. But as we found in Chapter 9, politics cannot be ignored in a democracy like ours. If the public believes strongly enough that solar is good and nuclear is bad, and that it is worth any price not to use nuclear power, we will not have nuclear power. Our country will suffer economically from competition with others that do use nuclear power, but that will only be part of the price we will pay. It is therefore extremely important that the public be properly educated before making such a decision.

Chapter 15 / QUESTIONS AND ANSWERS

Following my lectures on nuclear power, I normally get several questions. I shall address a representative sample of them here.

POWER NEEDS

Q. I understand that we have too many plants for generating electricity. Why, then, should we build more?

A. In the early 1980s, we had too many, but that has changed. From 1982-1989, U.S. electricity consumption increased by 30%. Some sections of the nation are already experiencing blackouts and brownouts, and the frequency of these will increase if our consumption continues to rise.

Insufficient electric power can stunt economic growth, and that can have many severe consequences on both our wealth and our health as long as our population is growing.

Q. Won't solar energy be taking care of our electric power needs?

A. While progress in developing photoelectric cells has been very encouraging, there is no prospect that they can provide a substantial fraction of our total electric power needs in the next 10-20 years.

Nuclear and solar power are not in competition, and probably never will be. Solar power cannot provide electricity at night. If we build nuclear power plants to provide nighttime power, it would be economically irrational not to operate them during the day, since their fuel costs are very low. Solar power will therefore be useful only to provide the additional power needed during the day.

This situation can change only if batteries for storing electricity become very much better and cheaper that those now available or under development.

PROBLEMS WITH FOSSIL FUELS

Q. How sure are we that the "greenhouse effect" is real?

A. Whether or not the recent succession of warm temperatures and droughts is due to the greenhouse effect is uncertain. But there is no question that, if we continue on our course of burning fossil fuels, it will eventually become very real. The only question is when. If we wait until we find out, it will be too late to take steps to avert it.

Q. How sure are we that acid rain is destroying forests and fish?

A. The evidence is not certain, but the scientific consensus is that there are substantial adverse effects. The public perception here is also important. For example, the political tension acid rain causes between Canada and the United States makes it worth spending money to reduce the problem. Spending this money also reduces air pollution.

Q. Can't we eliminate the air pollution from coal burning?

A. It is quite easy to eliminate the visible smoke, which consists of the large particles, but these are not responsible for the adverse health effects. There is technology, albeit expensive, for greatly reducing the sulfur dioxide emissions, but it is not clear that this would greatly reduce the health impacts of air pollution (although it would reduce acid rain). The problem is that we don't know which of the thousands of components of air pollution cause the most severe health effects. If we don't know what to eliminate, we can't eliminate it.

RADIOACTIVITY AND RADIATION

Q. Radioactivity can harm us by radiating us from sources outside our bodies, by being taken in with food or water, or by being inhaled into our lungs, but you seem to consider only one of these pathways. What about the others?

A. All of these pathways are treated in the scientific literature, but in order to simplify popular expositions it is usual to treat only the most threatening pathway.

Q. Cancers from radiation may take up to 50 years to develop, and genetic effects may not show up for a hundred years or more. How, then, can we say that there will be essentially no health effects from the Three Mile Island accident?

A. We have determined the radiation doses in millirems received by these people. From many experiences with people exposed to radiation, like the Japanese A-bomb survivors and patients treated for medical purposes, we can estimate the effects of each millirem of radiation. A wide variety of scientific studies, including theoretical modeling and experiments on animals and on cells in laboratory dishes, contribute to this estimation process.

Q. Measurements of radioactivity in air, for example, are made at a few monitoring stations. How do we know the levels may not be much higher at places where there are no monitoring stations?

A. Scientists choose locations of monitoring stations so as to minimize this possibility, making use of the considerable body of scientific information on how materials are dispersed under various weather conditions. This information also predicts relationships between readings at various stations, which are checked to give added confidence. Radiation levels at different locations can be predicted from the quantity of radioactivity released and a knowledge of the weather conditions including wind speed and direction, and temperature versus height above ground. The weather conditions around nuclear plants are constantly monitored. Much can be learned about a radioactivity release from measuring radioactivity on the ground surface up to several days later. In the Three Mile Island accident, air samples were collected by airplanes to give additional data, and photographic film from area stores was purchased and developed to measure the fogging by radiation (no fogging was observed, but it would have been if there had been appreciable radiation). There are thus many independent ways to determine the pattern of radiation exposure, and they serve as checks on one another.

This situation contrasts sharply with that for air pollution from coal burning. Since monitoring for air pollutants is much more difficult and expensive, there are very few monitoring stations even in a large metropolitan area. Nonetheless, air pollution kills many thousands of people every year and is thus a much greater threat to our health than is radioactivity from nuclear plants.

Q. Radioactive materials can be concentrated by various biological organisms. For example, strontium-90 ingested by a cow mostly gets into its milk. Doesn't this make radioactivity much more dangerous than your calculations indicate?

A. This is taken into account in all calculations and estimates. There was a widely publicized omission for the case of strontium-90 in milk in the 1950s, but that was a very long time ago, scientifically speaking.

Q. Air pollution may kill people now, but radiation induces genetic effects that will damage future generations. How can we justify our enjoying the benefits of nuclear energy while future generations bear the suffering from it?

A. Air pollution and chemicals released in coal burning also have genetic effects, as indicated by tests on microorganisms. While they are not as well understood and quantified, there is no reason to believe that the genetic impacts of coal burning are less severe than those of nuclear power.

The genetic impacts of radiation are not large. The total number of eventual genetic defects caused by a given radiation exposure added up over all future generations is less than half of the number of cancers it causes, less than one genetic defect for every 10 million millirem.

There are many ways in which our technology injures future generations, such as consuming the world's limited mineral resources, which will cause them infinitely more serious injury than genetic effects of our radiation. In the latter case, we are more than compensating our progeny with biomedical research that will greatly improve their health in many ways, including averting much of their genetic disease.

We leave future generations many legacies, both good and bad. The important point is that the good legacies outweigh the bad.

Q. Can the genetic effects of low-level radiation destroy the human race?

A. No. The process of natural selection causes good mutations to be bred in and bad mutations to be bred out. In the very long term, mutations from any given amount of radiation exposure disappear if they are harmful, and are preserved if they are beneficial; they can therefore only improve the human race, although that effect is negligibly small.

Humankind has always been exposed to radiation from natural sources, hundreds of times higher in average intensity than low-level radiation from the nuclear industry. Yet even this natural radiation is responsible for only a few percent of all genetic disease. Spontaneous mutations are responsible for the great majority of it.

Q. Isn't the artificial radioactivity created by the nuclear industry, to which man has never been exposed until recently, more dangerous than the natural radiation which has always been present?

A. The cancer and genetic effects of radiation are caused by a particle of radiation, say a gamma ray, knocking loose an electron from a certain molecule. In this process, there is no possible way, even in principle, for the electron or the molecule to "know" whether that gamma ray was originally emitted from a naturally radioactive atom, or from an atom that was made radioactive by nuclear technology. The answer to the question is NO.

Q. Can radiation exposure to parents cause children to be born with two heads or other such deformities?

A. NO. Such things are not occurring now although mankind has always been exposed to natural radiation. There is no possible way in which manmade radiation can cause problems that do not occur as a result of natural radiation. No new genetic diseases have been found among the Japanese A-bomb survivors or others exposed to high doses of radiation.

Q. Is there any factual basis for radiation creating monsters like "The Incredible Hulk"?

A. Absolutely none. These are strictly creations of an artist's imagination.

Q. You frequently use statistics to support your case. But it is well known that, "while statistics don't lie, liars can use statistics." How can we trust your statistics?

A. I make very little use of statistics, because there is no statistical evidence of harm to human health due to radiation from the nuclear industry. I use probabilities, which are something very different.* If you don't believe in probability, I would love to engage you in a game of coin flipping or dice rolling. Las Vegas, the various state lotteries, and insurance companies, are doing very well depending on the laws of probability.

One opponent of nuclear power has gotten lots of publicity by making flagrant use of statistics to deceive. If you look hard enough, you can always find an area around some nuclear plant that has a higher than average cancer rate. Of course, you can just as easily find one with a lower than average cancer rate, but he doesn't report that.

Q. How can you treat deaths due to radiation as statistics? These are human beings suffering and dying.

A. There are also human beings suffering and dying from air pollution, from chemical poisons, from poverty, and so on. Nuclear power will reduce these problems. I am only interested in reducing the total number of people who suffer and die.

Q. Isn't a nuclear accident that kills 1,000 people worse than having 10,000 people die one by one from air pollution with no one knowing why they died?

A. I thoroughly disagree. The only reason anyone can believe such a thing is because the media handle it that way. They would give tremendous publicity to a nuclear accident killing a thousand people (or even a hundred people), but they hardly mention the 30,000 or so people who die from air pollution every year from coal-burning power plants.

If you would choose a technology that kills 10,000 per year with air pollution over one that kills 1,000 in a large accident each year, you should be given the job of explaining to the extra 9,000 victims (and their loved ones) that they must die because people don't like media reports of large accidents. I'm sure you would quickly change your mind.

Q. Does radiation make people glow in the dark?

A. Radiation produces light only when extremely intense, as inside a nuclear reactor; people exposed to that much radiation would die instantly

---

*To illustrate the difference between probability and statistics, consider the honest flipping of a coin. The *probability* for heads is 50%. If you flip a coin 10 times and get eight heads, you could say that your *statistics* indicate that heads comes up 80% of the time.

(as they would inside any other furnace). Comedians get laughs from jokes about irradiated people glowing in the dark, but there is no factual basis for that idea.

Q. If we can't feel radiation, how do we know when we are being exposed?

A. There are numerous instruments for detecting radiation. Many of them are quite cheap, reliable, and sensitive. It is very much easier to detect radiation than the dangerous components in air pollution.

TRUST AND FAITH

Q. Why should we believe scientists when they have made nuclear bombs and all sorts of devastating weapons?

A. Working for the military is not an indication that a person does not tell the truth. Moreover, the scientists involved with nuclear power are an entirely different group (with a very few individual exceptions) than those who developed nuclear weapons.

I have never heard evidence that the scientific community as a group has deceived the public. Working as a scientist requires a high degree of honesty, if for no other reason than that dishonesty would be readily discovered and the career of an offender would be irreparably damaged by it.

Q. Since nuclear scientists rely on the nuclear industry for their livelihood, how can we believe them?

A. *University* scientists do not rely on the nuclear industry; in fact most of them have lifetime job security guaranteed by the university that employs them (I am in that position). Radiation health scientists would get *increased* importance and job security if people decided that radiation is *more* dangerous, because they are the ones who protect the public from radiation. If the nuclear industry were to shut down today, they would have a secure lifetime career from participating in the retirement of plants.

On the other hand, the few scientists who have vocally opposed nuclear power make a good living out of their opposition. They get large fees for speaking and for testifying in legal suits, and their books have sold well. If a nuclear scientist were interested in making extra money, he would do well to reverse his position and become an opponent of nuclear power.

A university nuclear scientist could become antinuclear without any repercussions on his job security. Scientists employed by organizations opposed to nuclear power, on the other hand, would instantly lose their jobs if they decided to become pronuclear.

Q. With nuclear scientists split on the question of dangers of radiation, how do we know which side to believe?

A. The split in the scientific community is not "down the middle" as the media would have you believe, but is very heavily one-sided. All of the official committees of prestigious scientists, like the National Academy of Sciences Committee, the United Nations Scientific Committee, the International Commission on Radiological Protection, the U.S. National Council on Radiation Protection and Measurements, the British National Radiological Protection Board, and similar groups in other countries, agree on the effects of radiation (within a range of differences that is irrelevant for purposes of public concern), and they are backed by the vast majority of the involved scientific community of specialists in radiation health. There has not been even a single vote by a single scientist on these committees supporting the views of people like Sternglass or Gofman which have been widely trumpeted by the media.

Q. Why should we trust the government when it gives us information on nuclear power?

A. The scientific evidence on health impacts of radiation has little dependence on government sources of information. The same is true of most other areas covered in this book. You are mainly being asked to trust the international scientific community.

Q. The nuclear "Establishment" told us that there could never be a reactor accident, but we had Three Mile Island. How can we trust them?

A. The nuclear "Establishment" did not say that there could never be a reactor accident. The Reactor Safety Study, which represents the nuclear Establishment more than anything else on that issue, estimates that there is a 30% chance that we would have had a complete meltdown by now (between civilian and naval reactors), whereas there has been none.

Q. How can we trust the nuclear Establishment when they construct nuclear power plants on earthquake faults?

A. The regulations on location of reactors relative to earthquake faults are very lengthy, complex, and technical. They have been worked out by some of the nation's foremost earthquake scientists. The distance from a fault allowed depends on the type of fault and the length of time since it has been active. Contrary to widely circulated stories, there is no reactor constructed on top of a fault or even within a mile of one. Earthquake scientists continue to do research on the matter. For example, after the severe 1989 earthquake in Armenia, they visited and carried out extensive investigations on the response of various structures. Small earthquakes, like the one near San Francisco in 1989, are of no consequence to nuclear plants.

Q. How can we trust reactor operators to do their job properly? How do we know they won't get drunk and cause an accident?

A. In safety analyses, it is not assumed that reactor operators are perfect; it is rather assumed that they make mistakes just like anyone else—pushing wrong buttons, failing to do required jobs, and the like. In accordance with the principle of defense in depth, the guiding philosophy in power plant design, there are back-up systems to compensate for such errors. The failure of any one link, of course, reduces the effectiveness of the defense in depth. It is therefore avoided as much as possible. Reactor operators must frequently pass stiff examinations; there are several operators as well as a graduate engineer on hand at all times; there are training programs, supervision, and inspections. But it is clearly recognized by all concerned that reactor operators are human beings and must be expected to behave as other human beings, which is a long way from perfect.

Q. How can we trust utilities not to take short-cuts in efforts to save money, thereby compromising safety?

A. Since I have no first-hand experience with utility construction practices, I cannot speak as an expert on this. But utilities are guaranteed a reasonable profit by their public utility commissions if they behave properly. They therefore have no great incentive to save money by cutting corners. Moreover, an accident in a nuclear plant is perhaps the most serious business blow a utility can suffer, costing its stockholders hundreds of millions or even billions of dollars. The utility that owns the Three Mile Island plant almost went bankrupt as a result of that accident.

In addition, the Nuclear Regulatory Commission (NRC) makes regular inspections. Anything not properly constructed may have to be torn out and reconstructed, at very great expense. A completed plant near Cincinnati had to be abandoned because it did not have documentation to prove that all welds had been properly inspected.

There is therefore a heavy incentive for utilities to behave properly in constructing plants. Similar considerations apply to operating plants. They are inspected frequently by the NRC, and heavy fines are levied for substandard practices. There are also careful inspections by the Institute for Nuclear Power Operations. This is an Atlanta-based organization sponsored by the nuclear industry because it recognizes that an accident in one plant causes difficulties for the whole industry.

REACTOR ACCIDENTS AND SAFETY

Q. Can a reactor explode like a nuclear bomb?

A. No, this would be impossible. A bomb requires fuel with more than 50% Uranium-235, whereas reactor fuel has only 3% Uranium-235. A bomb cannot work if the fuel is intermixed with water because the water slows down the neutrons, but the fuel in a nuclear power reactor is immersed in water. There are other reasons in addition that make a nuclear explosion impossible.

Q. Was the Three Mile Island accident a "close call" to disaster?

A. All studies agree that it was not, because there was no threat to the integrity of the containment.

Q. Could the Chernobyl accident happen here?

A. NO. The Chernobyl reactor was of an entirely different type than U.S. nuclear power reactors, and it had several features that made it much less safe.

Q. Is nuclear power safe?

A. Nothing in this world is perfectly safe. But in comparison with other methods available for generating electricity, or with the risks of doing without electricity, the dangers of nuclear power are very small. They are also hundreds of times smaller than many other risks we constantly live with and pay no attention to.

Q. Nuclear power is very new and different. How do you know that new and unsuspected problems will not develop?

A. There has been a tremendous research effort on areas of potential trouble, and commercial reactors have been operating for over 30 years. The U.S. Navy has been operating a large number of very similar reactors for about 30 years. But, of course, new and unsuspected problems have developed and probably will continue to develop as in any technology. That is why there are continuing research efforts, many avenues for information exchange among plants, and continued attention to the problems by the NRC. I see no reason to believe that we cannot keep ahead of the problems and maintain the present level of safety. As experience accumulates, we learn more about safety problems, thereby *improving* safety. Many valuable lessons were learned from the Three Mile Island accident. The new generation of reactors has benefitted from all of this research and experience, and will thereby provide a 1,000-fold improvement in safety.

Q. How can you trust the "fault tree analysis" method used in the Reactor Safety Study and other probabilistic risk analyses?

A. It is the best method available. If you don't trust it, you can fall back on experience. If its results were overoptimistic, we would have had more close calls by now than have actually been experienced. These analyses have been carried out independently by several different groups, including some in other countries, with similar results always being obtained. Of course, probabilistic risk analysis is an active area of science, with new ideas constantly being injected to improve the process.

Q. If reactors are so safe, why don't homeowners' insurance policies cover reactor accidents? Doesn't this mean that insurance companies have no confidence in them?

A. Insurance companies do insure reactors. In fact, they stand to lose more money from a nuclear accident than from any other readily conceivable mishap. However, they are limited by law in the amount they can insure against any one event, because if an insurance company were to fail, many innocent policy holders would be left without protection.

Liability insurance for reactors is covered by an act of Congress which provides over $7 billion of no-fault insurance on each plant, paid for by the utilities. Because of this insurance, coverage in home owners' insurance would provide double coverage and is therefore excluded. If the liabilities from an accident should exceed the maximum, Congress has stipulated that it will provide relief as is customary in disaster situations. Few other disasters are covered by as much private insurance. For example, there is little or no coverage for dam failures or bridge collapses.

Q. If reactors are safe, why are there evacuation plans for areas around them?

A. This is an example of regulatory ratcheting by the Nuclear Regulatory Commission. Until 1980, there were no such plans, and they are not used in other countries. There are no evacuation plans around chemical plants, although evacuations in their vicinity are more likely to be necessary than around nuclear plants. Most evacuations occur as a result of railroad or truck accidents involving toxic chemicals, but there is no advanced planning for them. It would be difficult to dispute the NRC viewpoint that having evacuation plans increases safety to some extent. They gave no consideration to the fact that the existence and advertising of these plans is unsettling to the public.

Radioactive Waste

Q. What are we going to do with the radioactive waste?

A. We are going to convert it into rocks, and put it in the natural habitat of rocks, deep underground.

Q. How do we know it will be safe?

A. We know a great deal about how rocks behave, and there is every reason to believe that waste converted to rock will behave the same. Calculations on that basis show that buried waste will be extremely safe. For example, it will do thousands of times less harm to public health than wastes from coal burning.

Q. If it's so easy, why aren't we burying waste now?

A. Research is being done to determine the *optimal* burial technology. There are important political reasons for doing this research before proceeding.

From an objective viewpoint, there need be no rush to bury the waste, and in fact there would be important advantages in storing it for 50-100 years before burial — 70-90% of the radioactivity and heat generation would decay away during this time period. However, because of the intense pressure generated by public misunderstanding, the U.S. government is forging ahead with plans to begin burying waste shortly after the turn of the century.

Q. Isn't disposal of radioactive waste an "unsolved problem"?

A. What is a solved problem? Some of the wastes from coal burning, better known as "air pollution," are simply spewed out into the air where they kill tens of thousands of Americans every year. Is that a solved problem? The solid wastes from coal burning are dumped in shallow landfills where, over the next hundred thousand years or so, they will kill many more thousands (see Chapter 11). Is this a solved problem?

We know many satisfactory solutions to the radioactive waste burial problem and are involved in deciding which of these is best. Any one of them would be thousands of times less damaging to health than our handling of wastes from coal burning.

Q. How long will the radioactive waste be hazardous?

A. It will lose 98% of its toxicity after about 200 years, by which time it will be no more toxic than some natural minerals in the ground. It will lose 99% of its *remaining* toxicity over the next 30,000 years, but it will still retain some toxicity for millions of years.

This situation is much more favorable than for some toxic chemical agents like mercury, arsenic, and cadmium, which retain their toxicity undiminished *forever*.

Q. How will we get rid of reactors when their useful life is over?

A. Fuel, which contains nearly all of the radioactivity, is removed every year. When a reactor is retired, the remaining fuel will be similarly removed and sent away for disposal as high-level waste. The residual equipment, which is only weakly radioactive, becomes low-level waste. Studies show that paying for this process adds less than 1% to the cost of electricity. The first commercial nuclear reactor at Shippingport, Pennsylvania, was disassembled and removed for $99 million, leaving the land available for unrestricted use.

MISCELLANEOUS TOPICS

Q. How long will our uranium supplies last?

A. With present reactors, we can continue building plants for about another 30 years and still be able to guarantee each a lifetime supply of fuel. Beyond that we will have to convert over to breeder reactors. With that technology, our fuel supply will last forever without affecting the price of electricity (see Chapter 13).

Q. Is nuclear power necessary?

A. The United States could get along for the foreseeable future with coal. It would be more expensive as well as much more harmful to our health and to the environment, but we could get by. For other countries, the situation is much less favorable. Western Europe and Japan have relatively little coal, and will therefore sorely need nuclear power when the oil runs out or is withdrawn for political or economic reasons. For them, nuclear power is much cheaper than any other alternatives.

Q. What harm could terrorists do if they took control of a nuclear power plant?

A. In principle, they could cause a very bad accident, thereby killing tens of thousands of people, including themselves. However, nearly all of their victims would suffer no immediate effects, but rather would die of cancer 10 to 50 years later. In view of the high normal incidence of cancer, these excess cases would be unnoticeable (see Chapter 6). This would hardly serve the purposes of terrorists.

By contrast, there are many simple ways these terrorists could kill at least as many people immediately. For example, they could put a poison gas into the ventilation system of a large building. Other examples are given in Chapter 13.

Nuclear power plants have very elaborate security measures with over a dozen armed guards on duty at all times, electronic aids for detecting intruders, emergency procedures, radio communication, and so on. To sabotage a nuclear plant effectively would require a considerable amount of technical knowledge, and a substantial quantity of explosives.

Q. Can reactors be converted into weapon factories?

A. If a reprocessing plant is available, the plutonium in power reactors is usable for weapons but is of very poor quality for that purpose (see Chapter 13). Under nearly all circumstances, a nation desiring nuclear weapons would find it much cheaper, faster, and easier to produce the plutonium in other ways. This would also give their bombs higher explosive power and much improved reliability.

Q. Could terrorists steal nuclear materials and use them to make a nuclear bomb?

A. There is no material in the present U.S. nuclear power industry that can be used for making nuclear bombs. If we should do reprocessing of nuclear fuel, plutonium would be separated. But there would be many very formidable problems faced by terrorists in trying to steal it and use it to make a bomb (see Chapter 13), and their bomb would be a small one, capable of destroying about one city block. Terrorists could do much more harm much more easily by using more conventional means.

Q. Why did the U.S. Government develop nuclear energy while ignoring solar energy?

A. All objective analyses done during the 1950s and 1960s indicated that generating electricity from nuclear energy should be several times cheaper than generating it from solar energy. Moreover, solar energy is available only when the sun shines, which greatly increases the problems and the costs if it is used as a major power source.

No one, at that time, could foresee the political opposition to nuclear power which has driven its cost so high in the United States.

Q. Is the cost of waste disposal and the cost of eventually decommissioning* the plant included in the cost estimates for nuclear power?

A. The cost of waste disposal represents only about 1% of the cost of nuclear electricity. A tax to cover this cost (with something to spare) is included in the consumer's electric bill, increasing it by about 1%. The cost of decommissioning is to be borne by the utility, so they include it in their calculation of the rate to charge customers. It contributes less than 1% to that rate.

Q. Your discussion is too technical for us to understand. How can you expect a nonscientist to follow your arguments?

A. I have done my best to make my arguments as understandable as possible. The one thing it is impossible to do, however, is to answer the criticisms of nuclear power without giving *quantitative* demonstrations of nuclear versus alternative technologies. Doing anything less than that, I could easily make any technology appear to be as dangerous, or as safe, as I choose.

If you are not willing to follow those quantitative demonstrations, you could just accept the results with the understanding that they have been accepted by the great majority of the scientific community. These results are stated most succinctly in Chapter 8 (results from opponents of nuclear power in parentheses): having all of our electricity generated by nuclear power plants would reduce our life expectancy by less than 1 hour (1.5 days), making it as dangerous as a regular smoker smoking one

---

*Decommissioning refers to taking a plant out of service, dismantling it, and restoring the site for other uses.

extra cigarette every 15 years (3 months), as an overweight person increasing his weight by 0.012 ounces (0.8 ounces), or as raising the U.S. highway speed limit from 55 to 55.006 (55.4) miles per hour, and it is 2,000 times (30 times) less risky than switching from midsize to small cars.

Chapter 16 / RECAPITULATION

Many areas of the United States are already short of electrical generating capacity. *Brownouts,* situations in which voltage has to be reduced causing lights to dim and motors to run slow, have been necessary in several situations. In other cases, utilities have had to appeal for reduced usage, such as abstaining from use of air conditioners and clothes dryers or asking that nonessential commercial operations be shut down. In December 1989 there were complete blackouts in sections of Florida and in Houston, Texas. The situation is rapidly getting worse as our electricity consumption increases much more rapidly than the generating capacity needed to provide for it. We will soon have no choice but to launch into a large program of new power plant construction.

With minor exceptions, these new plants will have to be powered by coal, oil, natural gas, or nuclear fuels. There are lots of good reasons for avoiding the use of oil and gas to generate electricity:

- They are substantially more expensive than coal or nuclear fuels.
- World supplies are quite limited on a long-term perspective.
- They are essentially our only option for providing transportation by land, sea, or air.
- They are vitally needed as feedstock for manufacture of plastics, organic chemicals, and other products essential for our technology.
- Paying for imported oil is a heavy strain on our national economy, and this problem is rapidly getting worse.
- Our oil supplies are vulnerable to being cut off for political reasons.
- Oil prices are susceptible to very large and rapid increases.
- Oil dependence can lead to war.

For the most part, therefore, our new electrical generating capacity must be powered by coal or nuclear fuels, although oil and gas will still be used to some degree. Burning coal, oil, and gas leads to a wide variety of environmental problems. They are major contributors to the greenhouse effect, which threatens to cause highly disruptive climate changes:

- Agriculture will suffer severe blows like an end to growing soybeans and corn in the South and corn and wheat in the Great Plains.
- Farmers will also have to deal with increased livestock disease, and heavy damage from insect pests.
- Forests will undergo stress, as some species of trees will die off and have to be replaced by others.
- Seacoast areas will be subject to flooding.
- Waterfowl and various types of aquatic life will be seriously affected by reduction in wetlands areas.
- Insect plagues, droughts, forest fires, tornadoes, and floods will increase.

Burning coal is the major contributor to acid rain which, in some areas, is heavily damaging forests and fish in lakes. This acid rain is straining relations between Canada and the United States, and between several pairs of European nations.

But perhaps the most serious environmental problem with burning fossil fuels is air pollution, which is estimated to be killing about 100,000 Americans every year. Attempts to solve this problem are very expensive, and there is little reason to be confident that the limited objectives these attempts target will solve it. Air pollution causes a variety of illnesses, and it has several other unpleasant aspects, such as foul odors and the degrading of all sorts of objects from stone carvings to clothing.

Coal burning causes many other environmental problems, such as destruction of land surfaces by strip-mining, acid mine drainage, which pollutes our rivers and streams, land subsidence, which damages and destroys buildings, and waste banks from washing coal, which are ugly and lead to air pollution. Coal mining is a harsh and unpleasant occupation. Miners are frequently killed in accidents, and constant exposure to coal dust causes severe degradation in their health, often leading to premature death from an assortment of lung diseases.

Oil has its environmental problems too. It contributes substantially to air pollution and to acid rain. Oil spills in our oceans have fouled beaches and caused severe damage to aquatic life. Oil causes fires, odors, and water pollution. The use of natural gas can lead to fires and explosions and can kill people through asphyxiation.

All of the adverse health and environmental effects resulting from burning coal, oil, or natural gas to produce electricity can be avoided by the use of nuclear power. As the public becomes more concerned about these problems, its attitude toward nuclear power is changing. Recent polls show that the American public now recognizes the need for new energy supplies, and that it wants and expects a much larger contribution from nuclear power. In fact, a substantial majority of the public believes that nuclear power will, and should, supplant coal as our primary source of electricity generation in the very near future. At the same time, the nuclear industry has been developing new types of power plants that are cheaper and very much safer than facilities now in operation. The stage seems to be set for a new surge in building nuclear power plants.

However, not far below the surface and threatening to erupt at any time is a large reservoir of concerns about the health and environmental impacts of nuclear power. Most of this book has been devoted to addressing these concerns. They derive from a number of misunderstandings, mostly large differences between the public's impressions and the viewpoints of the scientific community on important issues.

PUBLIC MISUNDERSTANDING

The most important misunderstanding concerns the hazards of radiation. The public views radiation as something quite new, highly mysterious, and extremely dangerous. Actually, there is nothing new about radiation; humans always have been, and always will be, exposed to radiation from natural sources, at hundreds of times higher levels than they will ever experience from the nuclear industry. Far from being mysterious, radiation is much simpler and better understood than air pollution, food additives, insecticides, or nearly any other environmental agent.

The *danger* from radiation is a quantitative issue, and thus it must be considered quantitatively. A very extensive basis is available for estimating the risks of radiation exposure. The principal source of information is experience with people exposed to high radiation doses from the atomic bomb attacks on Japan in World War II, from a wide variety of medical treatments with X-rays and radioactive sources, and from occupational exposures to radiation. We also have very extensive information from experiments on animals and on microorganisms in laboratory dishes. Studies of the health effects of radiation comprise a well-developed scientific field with strong interactions between theory and experiment, and with strong links to other fields of science. As a result, there is essentially unanimous agreement in the scientific community on the quantitative estimates of the dangers of radiation, or at least of its maximum danger. (Many believe that the dangers of low-level radiation like that received from use of nuclear power could even be zero.) These scientific estimates are presented in the reports of several national and international committees and commissions, consisting of our most distinguished scientists.

Unfortunately, the public has developed a very clear impression that the danger of radiation is incomparably greater than what is indicated by these scientific estimates. The reasons for this large difference include heavy media coverage of even the most trivial incidents involving radiation exposure, use of inflammatory adjectives like "deadly" or "lethal" in describing radiation, and frequent TV appearances by scientists from far outside the mainstream.

The second major misunderstanding concerns the frequency and consequences of a large reactor accident. The government agencies involved with nuclear power have sponsored very extensive research on these topics. They have produced reports which outline a spectrum of potential serious accidents and estimate their probabilities of occurrence. Unfortunately, the public has been informed about only the most serious of these potential accidents, without being informed about their extremely low probability. For example, the accident that has been described most often is expected only once in 10 million years. When the probabilities are correctly taken into account, it turns out that, averaged over time, we can expect less than 5 deaths per year from these accidents, as compared with tens of thousands of deaths per year from coal-burning air pollution. Consequently, every time a coal-burning plant is built instead of a nuclear plant, thousands of extra citizens are condemned to an early death. This statement applies even if we accept the risk estimates by the leading group opposed to nuclear power.

The accident at the Chernobyl nuclear power plant in the USSR stirred up old fears, and it was only natural to ask whether such an accident could occur here. However, that reactor was of a very different type than those used in the United States because it was designed to produce plutonium for nuclear bombs. This required several compromises on safety that have always been rejected as unacceptable in the West. The principal lesson learned from analyses of the Chernobyl accident is that our approach to reactor safety is far superior to that of the Soviets.

The third major misunderstanding is about the dangers of radioactive waste, principally the "high-level waste" produced directly in the fuel by the energy-generating reactions. This waste is often viewed as almost infinitely toxic. Actually, it contains far less toxicity than many other products of our technology like chlorine, ammonia, phosgene, barium, or arsenic compounds, or air pollution from burning coal.

What are we going to do with this waste? We're going to convert it into a rocklike material and bury it in the natural habitat of rocks, deep underground. How do we know it will be safe? We know all about how rocks behave, and there is every reason to believe that this material will behave similarly. Engineered features will be added to provide extra safety. For example, the waste packages will be sealed in corrosion-resistant casings that provide a high degree of safety even if all other protective features should fail.

The fact that radioactivity levels decrease with time eliminates nearly all of the danger, since movement of material to the surface is a very slow process. It is retarded by a succession of barriers: absence of groundwater, insolubility of the surrounding rock, sealing action of backfill material, corrosion resistance of the casing, insolubility of the waste itself, slow movement of groundwater (one inch per day in the site under investigation), filtering action of the rock, which constantly removes dissolved materials from the groundwater, and long distances for the groundwater to flow before it reaches the surface. As a result, it will take hundreds of thousands or millions of years for buried waste materials to get back into the environment. By that time, the radioactivity in the waste will be only a tiny fraction even of the radioactivity that was originally mined out of the ground to produce the fuel.

Quite aside from engineered safeguards and time delays, analyses based on a randomly selected burial site indicate that only 1 atom in 100 million of the buried waste will reach the surface in any one year, and only

1 atom in 10,000 of these will enter human bodies. Quantitative calculations then predict only 0.02 deaths over all future time due to the high-level waste produced by one nuclear plant in one year. This is a thousand times less than the effects of air pollution, one of the wastes from coal-burning power plants.

The U.S. Department of Energy has tentatively selected a site for the first repository in the Nevada desert, and has embarked on a program to evaluate its safety. All available information indicates that it will be very much safer than the randomly located repository considered above. Because of the public concern, hundreds of millions of dollars are being spent annually on this program. The costs are being covered (with lots to spare) by 0.1 cent per Kilowatt-hour tax on nuclear electricity. This still adds only 1% to the cost of nuclear power.

Aside from high-level waste, there are other radioactive waste problems. One that has attracted substantial attention is low-level waste from nuclear power plants. But the impact of this low-level waste on public health is much smaller, perhaps 2% of the already very small impacts of high-level waste. The most important waste by far is radon, released into the environment from mining uranium for nuclear fuel. Mining the uranium to fuel one nuclear power plant for one year will eventually result in releasing enough radon to cause 11 deaths. But conversely, removing the uranium from the ground will eventually save hundreds of lives that would otherwise be lost due to the radon it would generate if this uranium were left in the ground. Since radon is a gas, it naturally percolates up out of the ground where it can easily be inhaled by people.

When the same methods used to evaluate the dangers of buried nuclear wastes are applied to some of the wastes from coal-burning that end up in the ground, it turns out that two classes of this coal burning waste are many hundreds of times more harmful to public health than any of the nuclear wastes. One of these classes is the chemical carcinogens—cadmium, arsenic, and so forth; the other is radioactive materials—uranium, thorium, and radium—that eventually turn into cancer-causing radon.

The fourth area of public misunderstanding is about risks in our society. The public does not seem to realize that life is full of risks. Let us give a sample of them, along with the number of days by which they reduce the life expectancy of those exposed (an asterisk indicates the loss of life expectancy in days by the entire U.S. population): being poor, 3,500; smoking, 2,300; working as a coal miner, 1,100; being 15 pounds overweight, 450; motor vehicle accidents, 180*; using small rather than midsize automobiles, 60; radon in homes 35*; accidents with guns, 11*; using birth control pills, 5; hurricanes and tornadoes, 1*; living near a nuclear power plant, 0.4; large nuclear power program in the United States, 0.04*. Even according to the opponents of nuclear power, this last number is only 1.5.

It is clear from this list that nuclear power is a relatively trivial risk, and that its risks have received far too much attention. Another way to understand this fact is to consider how much money our society is willing to spend to save a life. Typical amounts are: programs in Third World nations, $200; cancer screening, $75,000; highway safety, $120,000; air pollution control, $1 million; natural radioactivity in drinking water, $5 million; nuclear power plant safety, $2.5 billion. In a democracy such as ours, government spending is determined by the degree of public concern. Hence, these numbers correspond roughly to the ratio of the danger perceived by the public to the true danger. It is clearly evident that the public's perceptions of the dangers of nuclear power are exaggerated at least a thousandfold.

Another public misunderstanding lies in the supposed connection between nuclear power and nuclear bombs. Actually this connection is very tenuous. Bombs made from plutonium produced in U.S.-type nuclear power plants would be of very inferior quality. Much easier and cheaper ways are available for a nation to procure much higher-quality nuclear weapons. The threat of terrorists stealing plutonium to make a bomb is greatly exaggerated; there are much easier and safer ways for terrorists to kill far more people if that is their goal. The toxicity of plutonium has also been highly exaggerated. Many things, like nerve gas or toxic biological agents, are far more deadly and much more easily dispersed into the environment, either by terrorists or through accidents.

COST PROBLEMS

Because of these misunderstandings and exaggerated concern about nuclear safety, regulatory requirements on nuclear power plants were constantly tightened in the late 1970s and early 1980s. This process required frequent design changes in the course of construction, which led to a great deal of wasted time and effort. As a consequence, the cost of a nuclear plant, corrected for inflation, quadrupled—dollar costs increased 10-fold. The effect was to make nuclear power economically unattractive. No nuclear power plant construction projects have been started since the mid-1970s, and many dozens of projects have been cancelled.

The basic problem was that nuclear power plants were not conceptually designed for the super-super safety that the public now demands. Achieving improved safety by add-on systems is both inefficient and limited in what can be achieved. The nuclear industry has therefore started over with new conceptual designs and is developing a new generation of reactors that will be a thousand times safer than those now in service. They will also be smaller and simpler, with far fewer things that can go wrong. One of their new features is passive stability; that is, even if electric power fails and the reactor operators simply walk away, no serious consequences occur. Electricity produced by these new nuclear reactors will be about 20% cheaper than that produced by coal-burning power plants, their principal competitor.

Some people advocate holding back on further use of nuclear power because they believe that solar electricity is "just around the corner." Actually, nuclear and solar electricity are not in competition, because the latter is not available at night. Even if all goes very well in the development of solar electricity, it can be useful only for providing the additional power needed during the daytime. Nuclear power is not normally used for that purpose. Any real competition between nuclear and solar electricity must therefore await the time when technology for storage batteries develops far beyond its present program goals.

As we face up to our growing need for more power plants, the only real choice in most cases is between nuclear and coal burning. Nuclear power will be substantially less expensive and thousands of times less harmful to our health and to our environment. The time truly seems to be ripe for a resurgence of nuclear power in the United States.

APPENDIXES

CHAPTER 3

More about the Greenhouse Effect[14]

Carbon dioxide is responsible for only about half of the Earth's greenhouse effect warming. Methane, which is increased by escaping natural gas, contributes 18%, and nitrogen oxides, which are products of fossil fuel burning, account for 6%. Another major contributor is chlorinated fluorocarbons (CFCs), which are responsible for 14%. These are familiar as the working material in air conditioners, refrigerators, and freezers, and as the propellant in spray cans. (Other important sources of methane include flooded rice paddies and cows burping up gas from their stomachs.)

Another cause of the greenhouse effect is cutting down forests, since plants take carbon dioxide out of the air. Clearing of the Amazon rainforest is believed to be especially serious. Worldwide, an area of forests equal to the area of the state of Virginia is cleared every year. This is estimated to cause 20% of the greenhouse effect.

Burning wood (or biomass) releases carbon dioxide, but it does not contribute to the greenhouse effect because the wood was created from carbon dioxide, which the tree leaves absorb from the atmosphere.

In 1950, the United States was responsible for 45% of the 1.6 billion tons of carbon dioxide emitted throughout the world, but by 1980 world emissions increased to 5.1 billion tons with only 27% from the United States; Western Europe's share was 23% in 1950 and 16.5% in 1980. Increased use of fossil fuels in underdeveloped countries has played an important role in increasing emissions.

CHAPTER 5

Genetic Diseases

We have referred to "genetic defects" and "genetic disease" several times in this chapter. For those interested we here present a deeper discussion of these.[11,25]

The male sperm and the female egg that unite in the reproductive process each carry 23 threadlike chromosomes, each of which is composed of many thousands of genes, and it is these genes that determine the traits of the newborn individual. At conception, the corresponding chromosomes from the two parents find one another and join, bringing the corresponding genes together. In the process, a fateful lottery takes place in which the traits of the new individual are determined by chance from a variety of possibilities passed down from previous generations. While the selections in each lottery are strictly a matter of chance, the overall process of genetics is not, because an individual with traits less favorable for survival is less likely to reproduce. This is Darwin's celebrated law of natural selection: favorable traits are bred into the race, while unfavorable traits are bred out.

But not quite all of the entries in the lottery are inherited from previous generations, because there can be changes in them, called *mutations.* The information contained in genes is encoded in the structure of the complex molecules of which they are composed, so if some change occurs in this structure, the information can be changed, leading to an alteration of the traits the genes determine. The great majority of these altered traits are harmful and are therefore referred to as *genetic defects.* Thousands of different diseases are medically recognized as arising from genetic defects. In fact, an appreciable fraction of ill health, with the notable exception of infectious diseases, is due to them. These genetically related diseases range from problems that are so mild as to be hardly noticeable to some so severe as to make life impossible. In the latter category, genetic defects are responsible for something like 20% of all spontaneous abortions.

It should not be inferred that every mutation causes a genetic defect. When the chromosomes from the sperm and egg join up at conception, there is a competition between each pair of matching genes to determine which determines the trait. Some genes are dominant and others are recessive, and in the competition, the former win out over the latter.

In a small fraction of mutations, the changed gene is dominant, causing its effects to be expressed in the child with high probability. For example, achondroplasia (short-limbed dwarfism), congenital cataracts and other eye diseases, and some types of muscular dystrophy and anemia are due to this type of mutation, and it sometimes causes children to be born with an extra finger or toe (which is easily removed by cosmetic surgery shortly after birth). About a thousand different medically recognized conditions are due to dominant mutations, and roughly 1% of the population suffers from them.

But in the great majority of cases, mutations are recessive, which means that they cause characteristic diseases only in the highly unusual situation where the same mutation occurs in the corresponding genes from both parents. Sickle cell anemia, cystic fibrosis, and Tay-Sachs disease are relatively well-known recessive diseases, but most of the 500-1,000 known recessive diseases are extremely rare, and only about one person in 1,000 suffers from them.

Another type of genetic damage involves the chromosomes rather than the genes of which they are composed. Chromosomes can be broken — recall that they are like threads — often followed by rejoining in other than their original configuration; for example, a broken-off piece of one chromosome can join to another chromosome. In some situations, there can be an entire extra chromosome or missing chromosome. About 1 person in 160 suffers from a disease caused by some type of chromosome aberration. Down's syndrome (mongolism) is perhaps the best-known example.

There is one type of genetic defect that is rather different from the ones we have been discussing. About 9% of all live-born children are seriously handicapped at some time during their lives by one of a variety of diseases that tend to "run in families," like diabetes, various forms of mental retardation, and epilepsy, in which genetic factors play a role but other factors are also important. It is estimated that the genetic component of these irregularly inherited diseases is something like 16%, with the other 84% due to environmental factors, food, smoking habits, and the like. Effectively, then,

16% of 9%, or 1.4%, of the population suffers from these diseases due to genetic factors. Adding them to the victims of dominant, recessive, and chromosomal disorders, we find that about 3% of the population is afflicted with some type of genetic disease.

Genetic Effects of Radiation

New mutations are constantly occurring in the sex cells of people destined to bear children, but this does not cause the 3% to increase, because mutations from earlier generations are being bred out by natural selection at an equal rate. That is, there is an equilibrium between introduction of new mutations and breeding out of old mutations. The great majority of new mutations occur spontaneously as a random process, due to the random motions that characterize all matter on an atomic scale. Occasionally these motions cause a complex molecule to break apart or change its structure, and that can result in a mutation. Similar damage can be done by foreign chemicals, or viruses, or radiation which happen to penetrate into the cell nucleus where the chromosomes are housed, but here we are concerned with the last of these, the genetic effects of radiation.

It should not be inferred that a particle of radiation striking the sex cells always leads to tragedy. On an average, the nucleus of every cell in our bodies is penetrated by a particle of radiation once every 3 years due just to the natural radiation to which we are all exposed, but this is responsible for only one-thirtieth as much genetic disease as is caused by spontaneous mutations (that is, ⅟₃₀ of 3%, or 0.1%, of the population suffers from genetic disease caused by natural radiation).

Some Calculations

In the introduction to Chapter 5, we discussed the number of gamma rays that strike an average person per second. To understand how these numbers are derived requires use of some definitions. One millirem (1 mrem) is defined for gamma rays as absorption by the body of 10^{-5} joules of energy per kilogram of weight. One MeV of energy is defined as 1.6×10^{-13} joules (J), and an average gamma ray from natural radiation has an energy of about 0.6 MeV, which is therefore $(0.6 \times 1.6 \times 10^{-13} =) 1 \times 10^{-13}$ J; 10^{-5} J then corresponds to 10^8 gamma rays. An average person weighs 70 kg (154 lb.), hence 1 mrem of radiation exposure corresponds to being struck by $(70 \times 10^8 =) 7 \times 10^9$ gamma rays. Natural radiation exposes us to about 80 mrem/year,[3] which corresponds to $(80 \times 7 \times 10^9 \approx) 5 \times 10^{11}$

gamma rays. There are 3.2×10^7 seconds in a year, so we are struck by $(5 \times 10^{11} \div 3.2 \times 10^7 \approx)$ 15,000 gamma rays every second. In a lifetime we are struck by about $(72 \times 5 \times 10^{11} \approx) 4 \times 10^{13} = 40$ trillion gamma rays. A medical X-ray may expose us to 50 mrem, which would be $(50 \times 7 \times 10^9 =)$ 3.5×10^{11} gamma rays. But the energy of X-rays is typically 10 times lower than the 0.6 MeV we have assumed for gamma rays, so ten times as many particles would be required to deposit the same energy; this is $(3.5 \times 10^{11} \times 10 =)$ 3.5 trillion X-ray particles striking us when we get an X-ray.

Since 1 mrem of gamma ray exposure corresponds to being struck by 7×10^9 particles, and the cancer risk from 1 mrem is 0.26×10^{-6}, the risk from being struck by one gamma ray is 1 chance in $(7 \times 10^9 \div 0.26 \times 10^{-6} \approx)$ 2.7×10^{16}, or 1 chance in 27 quadrillion.

Since natural radiation exposes us to 4×10^{13} gamma rays over a lifetime, the probability that it will cause a fatal cancer is $(4 \times 10^{13} \div 2.7 \times 10^{16} \approx) 1.5 \times 10^{-3}$, 1 chance in 700. Our total probability of dying from cancer is now 1 chance in 5, whence $(\frac{1}{700} \div \frac{1}{5} =)$ $\frac{1}{140}$ of all cancers are presumably caused by radiation.

CHAPTER 6

Probabilistic Risk Analysis

Probabilistic risk analysis, widely known as PRA, is the science of estimating the probability that some event will occur. The type of PRA used in analyzing reactor accidents, called "fault tree analysis," begins with identifying all "routes" leading to a meltdown. Each route consists of a succession of failures, like pipes cracking, pumps breaking down, valves sticking, operators pushing the wrong button, and so on. Since a given route will not lead to meltdown unless each of these failures occurs in turn, the probability of meltdown by that route is obtained by multiplying the probabilities for each individual failure. For example, if one particular route to meltdown consists of a pipe cracking badly—expected once in 1,000 years of operation—followed by a pump failing to operate—expected once in 100 trials—followed by a valve sticking closed—expected once in 200 attempts to open it, the chance that each of these three failures will occur successively in a given year is

$$\frac{1}{1000} \times \frac{1}{100} \times \frac{1}{200} = \frac{1}{20,000,000}$$

There has been extensive experience in many industries with pipes cracking, pumps failing, and valves sticking, so the probabilities for these are known (they depend on the quality of the pipes, pumps, and valves). It is these probabilities, obtained from experience, that are used in the calculations.

Once the probabilities for each route to meltdown are calculated, the probabilities for all possible routes must be added up to obtain the total probability for a meltdown. This is the largest source of uncertainty, since there is no way to be certain that all possible routes have been included. However, with many dozens of independent researchers thinking about these questions for many years, it seems reasonable to believe that at least most of the important routes have been considered.

The Emergency Core Cooling System—Preventing Core Damage[11]

Following a reactor shutdown, as in an accident, radioactivity in the fuel continues to generate heat at a rate shown by the curve[6] in Fig. 1. Ten seconds after shutdown, heat generation is at 5% of the full-power rate, and this drops to 3% after 3 minutes, 1% after 3 hours, and 0.3% after 5 days. If this heat is not carried away, the fuel will melt.

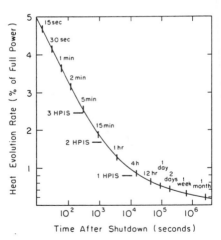

Fig. 1. The rate at which heat is evolved by the radioactivity in the fuel, as percentage of the heat evolution rate when the reactor is operating normally, versus time after the accident. The horizontal lines labeled "HPIS" show the amount of heat for which the water it boils off is equal to that supplied by high-pressure injection systems.

Under normal circumstances, the reactor fuel is submerged in rapidly flowing water, which picks up this heat and carries it to some other part of the plant where it is transferred to other systems. There are several things that can go wrong with this routine:

1. The other system may fail and be unable to accept this heat. (Such a failure—the breakdown of a pump in combination with valves on backup pumps being left closed for reasons still not explained—initiated the Three Mile Island accident, although that problem was quickly corrected and would have caused no trouble if it were not for other problems.)
2. The water flow may be blocked (as by a hydrogen bubble in the late stages of the Three Mile Island accident).
3. The water may escape slowly through a small leak or open valve. (This was the most important failure in the Three Mile Island accident; a valve failed to close and the operators did not recognize that fact.)
4. The system may burst open, releasing the pressure and thereby converting the water (which is at 600°F) into steam that would come shooting out through the opening—blowdown—leaving the reactor fuel with no water cooling.

In situations 1 and 2, the water will overheat and begin to boil, and the resulting steam will be released through pressure relief valves; these situations therefore also result in loss of water.

If the fuel is not covered with water, it will overheat and eventually melt; thus in all of these situations, it is important to inject more water into the reactor to replace that which is lost. In situations 1, 2, and 3, the reactor remains at high pressure (more than 100 times normal atmospheric pressure); it therefore requires special high-pressure pumps to inject water into it. There are typically three or four of these "high-pressure injection systems" (HPIS).[4,7] They provide enough water to make up for boil-off caused by the amounts of heat shown by the horizontal lines labelled "HPIS" in Fig. 1. As an illustration of their meaning, if three HPIS are working, enough water is injected to match the heat evolving from the fuel after 6 minutes; after that time, therefore, the total water in the system begins to increase. Up to that time, a calculation shows, only 2% of the water has been boiled away.

If only one HPIS is providing water, the rate at which water is provided is not equal to the rate at which it boils off until after 5.5 hours, according to Fig. 1. By that time, nearly 40% of the water would be boiled away, but there would still be more than enough left to keep the fuel covered. Thus any one of three or four HPIS would normally provide enough water to prevent a meltdown, or even damage to the fuel due to overheating. (In the Three Mile Island accident, the HPIS were turned off by the operators because they misinterpreted the information available to them as indicating that there was too much water in the reactor.)

In situation 4, the water is lost by blowdown in a matter of seconds; it is therefore important to get a lot of water back into the system immediately. This would be accomplished by systems called "accumulators"—large tanks filled with water at about one third the pressure in the reactor. The water in these tanks is normally kept out of the reactor by a valve which is held shut by the higher pressure from the reactor side. But if the pressure in the reactor should fall below that in the accumulator tanks, the latter would push the valve open, rapidly dumping the water from those tanks into the reactor. Note that this is a failproof system, not requiring electric power or any human action. There is enough water in two accumulators to keep the fuel covered for about 15 minutes before boil-off would lower its level below the top of the fuel.

Another element of the emergency core cooling system for adding water following a blowdown is two low-pressure injection systems (LPIS), either of which would provide enough water to cover the fuel in about 3 minutes, and enough to more than compensate for water boil-off at all times.[4,7] If either of these goes into operation by the time the water from the accumulator becomes insufficient, there can be no danger of damage to the fuel due to lack of water. But even if they fail, any one of the HPIS could provide sufficient water for this purpose. (They inject much more water when the reactor is at low pressure.)

Note that both the HPIS and the LPIS require electric power to drive pumps. If there should be an electric power failure following a blowdown, there would thus be about 15 minutes—the time during which water from the accumulators keeps the fuel covered—to restore electric power, as by starting up one of the diesel generators.

In summary, if even one of the three or four HPIS works, it is very difficult to imagine a situation in which insufficient water would be provided to prevent damage to the fuel. In the event of blowdown, either one of the LPIS or one of the HPIS would avert a meltdown.

The New PRAS[17]

In 1989, the Nuclear Regulatory Commission published a review of the state of the art PRAs on five U.S. nuclear power plants, three PWRs and two

BWRs, including the two originally evaluated in the RSS and three relatively new plants with very different types of containment. Only two of the five plants, one PWR and one BWR, were analyzed for earthquakes and fires. We will base the following discussion on that review, but the results of PRAs on other plants are similar.

The core damage probabilities due to internal failures (i.e., causes other than earthquakes and fires) in chances per million in one year is about 4 for the BWRs, 50 for two of the PWRs, and 340 for the other PWR. The reason for this last high risk is now recognized and is being corrected—this requires redesign of the reactor cooling water pump seal. For the two reactors where external events are included in the study, earthquakes add 40 chances per million for the BWR and 70 for the PWR, and fires add 20 for the BWR and 12 for the PWR. For both of the BWRs and one of the PWRs, the most important internal initiator is station blackout, although in one of the BWRs, the ATWS (anticipated transients without scram) accident is of nearly equal importance. In the two remaining PWRs, pipe-break LOCA is most important in one and pump seal failure LOCA is dominant in the other. For earthquakes, the most important initiator is loss of station power, but breaks in the high-pressure system are of nearly equal importance. For fires, station blackout is dominant in the BWR, and pump seal failure due to loss of its cooling water is most important in the PWR. Studies of earthquakes and fires in other reactors than the five considered here lead to similar conclusions.

The RSS studied two reactors, and both of them are also included here. For these, the core damage probability in the new studies, including earthquakes and fires which were not included in the RSS, is about the same as the meltdown probability in the RSS. For the internal failures it considered, the RSS was overconservative by a factor of 5 for the PWR and by a factor of 20 for the BWR, but its lack of consideration of earthquakes and fires makes up for this difference.

While the above figures give the impression that BWRs are much safer than PWRs, this is partly compensated by better containment performance for the latter. The probability for early containment failure and bypass, which are necessary conditions for appreciable health impacts, is 2% in the PWR with the high risk of core damage and 15% in the other two PWRs, versus about 50% in the BWRs. In core damage initiated by fire and earthquakes, these probabilities are about 3% and 10%, respectively, in the PWR versus 65% and 90% in the BWR.

The bottom line for these analyses is the number of deaths expected per year of operation. The number of deaths is the product of the probability for a core damage incident times the average number of deaths from such an accident. For internal initiators, the results are 0.015 deaths per year from the PWRs and 0.003 from the BWRs. The RSS gave about 0.04 for both. For fires, they are 0.0003 for the PWRs and 0.04 for the BWRs. No estimates were developed for earthquakes, because an earthquake that would cause core damage is one that would destroy dams, bridges, buildings, and roads, which would affect evacuation plans and sheltering from buildings. An alternative approach is to calculate by what percentage the nuclear plant releases would increase the total death rate from the earthquake. Best current estimates are 0.1%.

Nearly all of the above deaths would be from cancer many years later and could not be related to the accident. The average number of early deaths, within weeks following the accident, are 2, 25, and 70 millionths per year for the PWRs and 0.01 and 0.03 millionths per year for the BWRs. In the RSS they were 30 and 10 millionths per year for the PWR and BWR respectively.

Protections against Containment Rupture[11]

In nearly all scenarios for serious reactor accidents, a great deal of water ends up in the containment building. When heat is added to water, the latter can be converted into steam, which increases the pressure inside. If this pressure exceeds the maximum that the containment can withstand, the containment will rupture, allowing the release of radioactive material into the environment. The integrity of the containment thus depends on keeping the net heat evolved within the containment below the maximum allowable quantity.

This allowable amount in the absence of containment cooling is shown in Fig. 2 by the curve labeled "no cooling"; it increases with time because some heat is diffusing into the concrete walls, where it does not contribute to increasing the steam pressure. The sources of heat are shown by the dashed lines in Fig 2. The one source present in all cases is that generated by the radioactivity in the fuel. Its rate of evolution was shown in Fig. 1, and from that curve it is straightforward to calculate the total heat evolved up to any time; that is the dashed curve labeled "radioact" in Fig. 2. If blowdown occurs, the energy from it adds to the total, giving the dashed curve labeled " + blowdown." If the emergency core cooling system fails to restore cooling, there will be a chemical reaction between the fuel-cladding material, zirconium, and steam, which releases additional heat, bringing the total up to that shown by " + Zr-Steam." That reaction generates hydrogen, and if

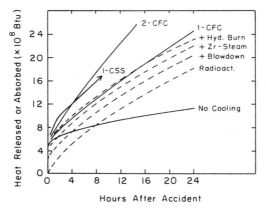

Fig. 2. The total quantity of heat released into the containment (dashed lines) and allowable without rupturing the containment due to high pressure (solid lines), versus time after the accident. The various lines are explained in the text.

this hydrogen burns, it contributes additional energy, bringing the total to the dashed curve labeled " + Hyd. Burn."

From Fig. 2 we see that the total energy released exceeds the allowable quantity with no cooling after 8 hours, even if only the radioactivity contributes, after 4 hours if there is a blowdown, and much sooner if the fuel overheats enough to allow the Zr-Steam reaction. Clearly, it is necessary to have systems for cooling the containment atmosphere.

There are two types of systems for doing this.[4] One of these is the containment spray system (CSS), which sprays water into the air to condense the steam. There are typically three of these. They would exhaust their stored water supply after 1-2 hours, leaving the water on the containment floor. The system would then be switched to pick up water from the containment floor, pass it through pipes surrounded by cool water from a separate source outside the containment, and then spray it into the containment air to achieve cooling. If one of these three systems is operating, there can be no danger of breaking the containment due to excess pressure inside. This can be seen from the curve labeled " + 1-CSS " in Fig. 2 which shows that the heat removed exceeds that provided by all sources combined by a large margin.

Another provision for containment cooling in some power plants is "containment fan coolers" (CFC), which blow air from the containment atmosphere over pipes carrying cooling water from a separate source outside the containment. There are typically five CFC, and the amount to which one or two of these increases the allowable amount of heat input is shown by the curves labeled "1-CFC" and "2-CFC" in Fig. 2. We see that if only one CFC is working, the cooling appears to be slightly insufficient to prevent containment rupture for the first 14 hours if all of the heat inputs contribute fully. However, if two of the five CFC are working, there is no problem with containment failure due to excess pressure.

CHAPTER 8

For those readers who are interested, we demonstrate here how to calculate some of the results quoted in Chapter 8.

First we calculate the LLE from reactor accidents according to the Nuclear Regulatory Commission Study which estimates one meltdown per 20,000 reactor-years of operation, and an average of 400 fatalities per meltdown. All base-load U.S. electricity derived from nuclear power plants would require about 250 such plants, giving us 250 reactor-years of operation each year. We would therefore expect a meltdown every $(20,000/250 =)$ 80 years on an average. The average fatality rate is then $(400/80 =)$ 5 per year. If the United States were to maintain its present population for a long time, there would be about 3 million deaths each year, so $(5/3 \text{ million} =)$ 1.7 out of every million deaths would be due to nuclear accidents. Victims of nuclear accidents lose an average of 20 years of life expectancy (cancers from radiation usually develop 10 to 50 years after exposure), giving the average American an LLE $= (1.7 \times 10^{-6} \times 20 =)$ 34×10^{-6} years; multiplying this by 365 days/yr $\times$ 24 hrs/day $\times$ 60 minutes/hr gives an LLE of 18 minutes, as quoted previously.

The UCS estimates are one meltdown every 2,000 reactor-years, with an average of 5,000 deaths per meltdown. Since these numbers are respectively 10 and 12.5 times higher than the NRC estimates, the LLE is larger by a factor of $(10 \times 12.5 =)$ 125. Multiplying this by 18 minutes gives 2,250 minutes, or 1.5 days. Alternatively, one could go through the entire calculation in the previous paragraph.

We next calculate the LLE from being exposed to 0.2 mrem/year, which is a lifetime exposure of $(70 \text{ years} \times 0.2 \text{ mrem/year} =)$ 14 mrem. In Chapter 5 we found the cancer risk to be 260×10^{-9} per mrem. Multiplying this by 14 mrem gives a lifetime cancer risk of 3.6×10^{-6}. Since the average victim loses about 20 years of life expectancy, the LLE is $(20 \text{ years} \times 3.6 \times 10^{-6} =)$ 7×10^{-5} years, or 37 minutes.

This corresponds to an LLE of 2.1 minutes per mrem (30 minutes/14 mrem). The average exposure in the Three Mile Island accident to people living in that area was 1.2 mrem, so their LLE was $(1.2 \times 2.1 =)$ 2.5 minutes.

The comparisons of risks are based on the ratio of LLE. For example, if 1 pound of added weight gives an LLE of 30 days while the UCS estimate gives an LLE from reactor accidents of 1.5 days, gaining 1 pound is $(30 \div 1.5 =)$ 20 times more dangerous, or gaining $\frac{1}{20}$ of a pound must be equally dangerous. Multiplying by 16 ounces per pound gives 0.8 ounces as the weight gain giving equivalent risk.

As an example of calculating the cost per life saved, consider the use of air bags. According to Allstate Insurance Company, an air bag reduces the driver's mortality rate by 1.4 deaths per hundred million miles driven. Therefore, if a car is driven 50,000 miles, its probability of saving a life is $1.4 \times (50,000/100,000,000) = 1/1,500$, or one chance in 1,500. This air bag would cost about $400, so the cost per life saved is $400 divided by 1/1,500, or $600,000. Another way of saying this is that for every 1,500 cars equipped with an air bag, an average of one life would be saved; the cost would then be $(1,500 \times \$400 =)$ $600,000 to save one life.

As another example, consider the Nuclear Regulatory Commission regulation requiring installation of any equipment in a nuclear plant that will reduce the total exposure to all members of the public by 1 mrem per dollar spent. It was shown above that we can expect one fatal cancer for every 4 million mrem; thus the regulation requires spending $4 million for each death averted.

CHAPTER 11

Probabilities for Entering the Human Body[10]

We inhale about 20 cubic meters (m^3) of air per day, or 7,000 m^3 per year. Dust levels in air from materials on the ground becoming suspended are about 35×10^{-6} g/m^3. Thus we inhale $(20 \times 35 \times 10^{-6} =)$ 0.7×10^{-3} grams per day of material from the ground, or 0.25 grams per year.

The area of the United States is about 10^{13} m^2, so the volume of the top inch (0.025 m) of soil is $(.025 \times 10^{13} =)$ 2.5×10^{11} m^3. Since the density of soil is 2×10^6 g/m^3, this soil weighs $(2 \times 10^6 \times 2.5 \times 10^{11} =)$ 5×10^{17} grams. Since each person inhales 0.25 g/yr of this soil, the quantity inhaled by the U.S. population (240×10^6) is $(240 \times 10^6 \times 0.25 =)$ 6×10^7 g/yr. The probability for any one atom in the top inch of U.S. soil to be inhaled by a human in one year is therefore $(6 \times 10^7/5 \times 10^{17} =)$ 1.2×10^{-10}, a little more than 1 chance in 10 billion.

The probability for an atom in a river to enter a human is very much larger. The total annual water flow in U.S. rivers is 1.5×10^{15} liters, whereas the total amount ingested by humans is 2.2 liters/person per day $\times$ 365 days/year $\times 240 \times 10^6 = 1.8 \times 10^{11}$ liters per year. Thus the probability for an atom in a river to be ingested by a human is $(1.8 \times 10^{11}/1.5 \times 10^{15} =) 1.2 \times 10^{-4}$, or a little more than 1 chance in 10,000 per year.

Radiation Reaching the Surface from Buried Waste

Rock and soil attenuate gamma rays, the most penetrating radiation from radioactive waste, by more than a factor of 3 per foot. Since the waste will be buried 2,000 feet underground, gamma rays from it would be attenuated by a factor of $3^{2000} \approx 10^{1000}$, an enormously large number. The radioactive waste produced by a million years of all-nuclear power in the United States would eventually emit about 10^{36} gamma rays;[11] thus the probability of even a single gamma ray from it reaching the surface would be about $(10^{36}/10^{1000} =) 10^{-964}$, an infinitesimally small probability (964 zeros after the decimal point).

Derivation[8] of Fig.1, Chapter 11

For illustrative purposes, let us calculate the number of liver cancers expected from people eating 1 millicurie (3.7×10^7 radioactive decays per second) of plutonium-239 (^{239}Pu) that is present in the waste.[2] Since the radiation emitted by ^{239}Pu, called alpha particles, does not go very far—it can barely get through a thin sheet of paper—this material can cause liver cancer only if it gets into the liver. Experiments indicate that 0.01% of ingested ^{239}Pu gets through the walls of the gastrointestinal tract into the bloodstream, and of this, 45% is deposited in the liver; thus $(3.7 \times 10^7 \times 0.0001 \times 0.45 =)$ 1,700 alpha particles strike the liver each second. Since ^{239}Pu remains in the liver for an average of 40 years (1.2×10^9 seconds), the total number of alpha particles that eventually strike the liver is $(1,700 \times 1.2 \times 10^9 =) 2 \times 10^{12}$. This is multiplied by the energy of the alpha particle to give the energy deposited, 1.5 joules, which is then divided by the mass of the liver, 1.8 kg, and multiplied by a conversion factor to give the dose in millirem, 1.6×10^6. The risk of liver cancer per millirem of alpha particle bombardment is estimated from studies of patients exposed for medical purposes to be 0.015×10^{-6} per millirem. The number of liver cancers expected from eating 1 millicurie of ^{239}Pu is therefore $(1.6 \times 10^6 \times .015 \times 10^6 =) 0.024$. Since the waste produced in one year by one plant contains 6×10^4 millicuries of ^{239}Pu, the number of liver cancers expected if this were fed to people would be $(6 \times 10^4 \times 0.024 =)$ 1,400.

But once ^{239}Pu gets into the bloodstream, it can also get into the bone — 45% accumulates there, and it stays for the remainder of life; it therefore can cause bone cancer. A calculation like that outlined above indicates that 700 cases are expected. When other body organs are treated similarly, the total number of cancers expected totals 2,300, which is the value of the curve labeled "^{239}Pu" in Fig. 1 (Chapter 11) in early time periods.

The quantity of ^{239}Pu in the waste does not stay constant, for two reasons. Every time a particle of radiation is emitted, a ^{239}Pu atom is destroyed, causing the quantity to decrease. But ^{239}Pu is the residue formed when another radioactive atom, americium-243, emits radiation, which adds to the quantity. When these two effects are combined, the quantity versus time is as shown by the curve in Fig. 1.

There are many other radioactive species besides ^{239}Pu in the waste; similar curves are calculated for them and shown in Fig. 1 (Chapter 11). The total number of cancers is then the sum of the number caused by each species, which is obtained by adding the curves at each time. The result is the thick curve above all the others in Fig.1 (Chapter 11). That is the curve used in our discussions as the number of cancers expected if all the waste produced by one power plant in one year were fed to people in digestible form.

Probability for an Atom of Rock to Enter a Human Stomach[15,19]

From the measured rate at which rivers carry dissolved material into oceans, it is straightforward to calculate that an average of 1.4×10^{-5} meters of depth is eroded away each year. Hydrologists estimate that 26% of this erosion is from dissolution of rock by groundwater; the rest is from surface water. Thus $(0.26 \times 1.4 \times 10^{-5} =)$ 3.6×10^{-6} meters of depth are dissolved annually by groundwater. The fraction of this derived from 1 meter of depth at 600 meters below the surface may be estimated from our knowledge of how groundwater flow varies with depth; it is about 2.6×10^{-4}. The total amount of rock derived from our 1 meter of depth in 1 year is then $(3.6 \times 10^{-6} \times 2.6 \times 10^{-4} =)$ 1×10^{-9} meters per year. If 1×10^{-9} meters is removed from 1 meter of depth each year, the probability for any one atom to be removed is 1×10^{-9}, one chance in a billion per year. This result applies to all rock, averaged over the continent.

We now provide an alternative derivation for an atom of rock which is submerged in groundwater. Consider a flow of groundwater, called an aquifer, along a path through average rock and eventually into a river. There is a great deal of information available on aquifers, like their paths through the rocks, the amount of water they carry into rivers each year, and the amounts of various materials dissolved in them. From the latter two pieces of information, we can calculate the quantity of each chemical element carried into the river each year by an average aquifer—i.e., how much iron, how much uranium, how much aluminum, and so on.

Where did this iron, uranium, and aluminum in the groundwater come from? Clearly, this material was dissolved out of the rock. From our knowledge of the path of the aquifer through the rock and the chemical composition of rock, we know the quantity of each of the chemical elements that is contained in the rock traversed by the aquifer. We can therefore calculate the fraction of each element in the rock that is dissolved out and carried into the river each year. For example, a particular aquifer may carry 0.003 pounds of uranium into a river each year that it dissolved out of a 50-mile-long path through 200 million tons of rock that contains 1,000,000 pounds of uranium as an impurity (this is typical of the amount of uranium in ordinary rock). The fraction of the uranium removed each year is then 0.003/1,000,000, or 3 parts per billion. Similar calculations give 0.3 parts per billion for iron, 20 parts per billion for calcium, 7 parts per billion for potassium, and 10 parts per billion for magnesium. To simplify our discussion, let us say that 10 parts per billion of everything is removed each year—this is faster than the actual removal rates for most elements. This means that the probability for any atom to be removed is 10 chances in a billion each year. This is 10 times larger than our first estimate, 1 chance in a billion. This is partly explained by the fact that most rock is *not* submerged in groundwater, as we have assumed here. But the important point is that we have given two entirely independent derivations and have arrived at roughly the same result. To be conservative, we will use the higher probability, 10 chances in a billion per year. Incidentally, this result implies that for ordinary rock submerged in groundwater, only 1% is removed per million years [(10/1 billion) × 1 million = 0.01], so it will typically last for 100 million years.

It was shown in the first section of this Appendix that an average molecule of water in a river has 1 chance in 10,000 of entering a human stomach before flowing into the oceans. For materials dissolved in the water, the probability is somewhat smaller because some of it is removed in drinking water purification processes, but the probability is increased by the fact that some material in rivers finds its way into food and enters human stomachs by that route. These two effects roughly compensate one another. We therefore estimate that material dissolved in rivers has 1 chance in 10,000 of getting into a human stomach.

Since an atom of the rock has 10 chances in a billion of reaching a river each year, and once in a river has 1 chance in 10,000 of reaching a human stomach, the overall probability for an atom of the rock to reach a human stomach is the product of these numbers [(10/1 billion) × (1/10,000) =], 1 chance in a trillion per year. That is the value used in our discussion.

In considering effects of erosion, we assumed that all the waste would be released into rivers after 13 million years. According to Fig. 1 in Chapter 11, if all the waste remaining at that time were to get into human stomachs, about 40 deaths would be expected. But we have just shown that if the material is released into rivers, only 1 atom in 10,000 reaches a human stomach; we thus expect (40/10,000 =) 0.004 deaths. That is the result used in our discussion.

CHAPTER 12

We have stated that by far the most important radiation health effect of nuclear power is the lives saved by mining uranium out of the ground, thereby reducing the exposure of future generations to radon. Let us trace through the process by which this is calculated.[2]

The radon to which we are now exposed comes from the uranium and its decay products in the top 1 meter of U.S. soil, since anything percolating up from deeper regions will decay before reaching the surface. From the quantity of uranium in soil (2.7 parts per million) and the land area of the United States (contiguous 48 states), it is straightforward to calculate that there are 66 million tons of uranium in the top meter of U.S. soil. This is now causing something like 10,000 deaths per year from radon, and will continue to do so for about 22,000 years, the time before it erodes away. This is a total of (10,000 × 22,000 =) 220 million deaths caused by 66 million tons of uranium, or 3.3 deaths per ton. As erosion continues, all uranium in the ground will eventually have its 22,000 years in the top meter of U.S. soil and will hence cause 3.3 deaths per ton.

In obtaining fuel for one nuclear power plant to operate for 1 year, 180 tons of uranium is mined out of the ground. This action may therefore be expected to avert (180 × 3.3 =) 600 deaths. However, to be consistent we must also consider erosion of the mill tailings covers. When this is taken into account, the net long-term effect of mining and milling is to save 420 lives.

On the short-term, 500-year perspective, we can ignore erosion, so uranium deep underground has no effect. However, about half of our uranium ore is surface mined, and about 1% of this $(0.005 \times 180 =) 0.9$ tons is taken from the top 1 meter. This saves $(0.9 \times 3.6 =) 3.3$ lives over the next 22,000 years, or $(500/22,000 \times 3.3 =) 0.07$ lives over the next 500 years. We must subtract from this the lives lost by radon emission through the covered mill tailings, 0.003/year without the cover, $\times \frac{1}{20}$ to account for attenuation by the cover, $\times 500$ years $= 0.07$. Thus the net effect over the next 500 years of mining and milling uranium to fuel one power plant for 1 year is to save $(0.07 - 0.07 =) 0.00$ lives. That is the result used in Table 1 in Chapter 12.

We next consider the effects of radon from coal burning. Coal contains an average of 1.0 parts per million uranium—some commercial coals contain up to 40 parts per million. When the coal is burned, by one way or another this uranium is released and settles into the top layers of the ground, where it will eventually cause 3.3 deaths per ton with its radon emissions, as shown above. The 3.3 million tons of coal burned each year by a large power plant releases 3.3 tons of uranium, which is then expected to cause 3.3 deaths per ton, or a total of 11 deaths. These will be distributed over about 100,000 years, with some tendency for more of them to occur earlier; thus about 1% of them, or 0.11 deaths, may be expected in the next 500 years. That is the figure in Table 1, Chapter 12.

On a multimillion-year time perspective, the uranium in the coal would have eventually reached the surface by erosion even if the coal had not been mined; these 11 deaths would therefore have occurred eventually anyhow, and hence can be discounted. However, the carbon, the main constituent of coal, is burned away, whereas if the coal had been left in the ground and eventually reached the surface by erosion, this carbon would not have emitted radon since it contains no uranium (the uranium in the coal has already been accounted for). Since that carbon is missing, its time near the surface will be taken by other rock which does contain uranium, 2.7 parts per million on an average, or $(3.3 \times 2.7 =) 9$ tons in the 3.3 million tons of rock which replaces the carbon. Multiplying this by 3.3 deaths per ton of uranium gives 30 deaths, the result we have used in Table 1, Chapter 12.

CHAPTER 13

Fraction of Cost of Nuclear Power due to Raw Fuel

Nuclear fuel undergoing fission produces 33×10^6 kW-days[45] of heat energy per metric ton (2,200 lbs). It requires about 6 tons of uranium to make 1 ton of fuel,* and one-third of the heat energy is converted into electricity.[45] Therefore, the electrical energy per pound of uranium is

$$\frac{\frac{1}{3} \times 33 \times 10^6 \text{ kW-days}}{6 \times 2200 \text{ lb}} \times \frac{24 \text{ hours}}{\text{day}} = 20,000 \frac{\text{kW-hr}}{\text{lb}}$$

Nuclear electricity costs about 5 cents/kW-hr, so the electricity production from one pound of uranium costs ($.05 \times 20,000 =$) \$1,000. Uranium costs about \$25/lb, which is .5% of this cost.

Cost of Gasoline versus Cost of Nuclear Fuel

From straightforward energy conversions, 1 pound of nuclear fuel undergoing fission is equivalent to 2.5×10^5 gallons of gasoline. Since the energy conversion efficiency in a breeder reactor[45] is 40%, the electrical energy from 1 pound of uranium, which costs about \$25, is equal to that in 1.0×10^5 gallons. The equivalent cost of gasoline is therefore (\$25/1 $\times 10^5$ gallon =) .025 cents/gallon. This is 40 gallons for a penny.

Present reactors burn only 1% of the uranium and are only 33% efficient, so the fuel cost is higher by a factor of ($100 \times 40/33 =$) 120. This is equivalent to gasoline costing ($120 \times .025 =$) 3 cents/gallon.

Radiation Dose to Lung from Plutonium and the Lung Cancer Risk

This calculation requires knowing (or accepting) some scientific definitions and may therefore not be understandable to many readers.

We calculate the dose to the lung from a trillionth of a pound of plutonium residing there for 2 years. The number of plutonium atoms is

$$10^{-12} \text{ lb} \times 450 \text{ g/lb} \times (6 \times 10^{23} \text{ atoms/239 gm}) = 1.1 \times 10^{12}$$

where 6×10^{23} is Avogadro's number and 239 is the atomic weight.[45] Since half of the plutonium atoms will decay in 24,000 years (the half life), the fraction undergoing decay during the 2 years it spends in the lung is a little more than one in 24,000; actually it is 1/17,000. The number that decay is then ($1.1 \times 10^{12}/17,000 =$) 7×10^7. Each decay releases an energy of about (5 MeV $\times 1.6 \times 10^{-13}$ joules/MeV =) 8×10^{-13} joules, so the total radiation energy deposited is ($7 \times 10^7 \times 8 \times 10^{-13} =$) 5.6×10^{-5} joules.[45] The

*In isotopic enrichment, 1 lb of natural uranium input contains 0.007 lb U-235, while the depleted product contains 0.002 lb; thus it contributes 0.005 lb of U-235 to the enriched product. One pound of reactor fuel contains 0.03 lb U-235 from this enriched product; it therefore requires (0.03 ÷ 0.005 =) 6 lb of uranium input.

weight of the average person's lung is 0.57 kg[46]; thus the energy deposited is $(5.6 \times 10^{-5} \div .57 =) 1 \times 10^{-4}$ joules/kg. The definition of a millirad[45] is 1×10^{-5} joules of energy deposit per kg of tissue.* The dose is therefore $(1 \times 10^{-4} \div 1 \times 10^{-5} =)$ 10 millirad. Since only 15% of what is inhaled spends this 2 years in the lung,[23] the exposure per trillionth of a pound *inhaled* is $(10 \times 0.15 =)$ 1.5 millirad. For alpha particles—the radiation emitted by plutonium—1 millirad equals 20 millirem,[47] so the dose to the lung is $(1.5 \times 20 =)$ 30 millirem per trillionth of a pound inhaled.

Estimates by BEIR,[24] UNSCEAR,[25] and ICRP[47] give a risk of about 5×10^{-7} lung cancers per millirad of alpha particle exposure. The number of lung cancers per pound inhaled is therefore $(1.5 \times 10^{12} \times 5 \times 10^{-7} =)$ 8×10^5. Mays[27] estimates 4×10^5 liver and bone cancers per pound inhaled, bringing the total effect to 1.2 million cancers of all types per pound inhaled.

There are two factors modifying this estimate. One is that this calculation is for young adults; averaging over all ages reduces the risk by half. The other factor is due to the fact that our calculation was for Pu-239, whereas typical samples of plutonium contain a mixture of other isotopes that generally contain more radioactivity per pound because they have shorter half lives.

When these factors are taken into account,[26] the deaths per pound inhaled become 4.2 million for wastes from present reactors, 2.7 million for breeder reactor fuel, and 0.8 million for weapons plutonium. In this chapter we have used 2 million deaths per pound as a loose average, mainly because that number has been used in most studies whose results are quoted here. All results quoted can be adjusted by taking effects to be proportional to these numbers.

---

*Millirad is the unit of physical radiation exposure, as indicated by this definition, whereas millirem includes a correction for biological effectiveness. For X-rays, beta rays, and gamma rays, 1 millirad is equal to 1 millirem, but for alpha particles, 1 millirad is equal to 20 millirems.

REFERENCES

CHAPTER 1

1. E. Sternglass, *Secret Fallout: Low Level Radiation from Hiroshima to Three Mile Island* (McGraw-Hill, New York, 1981).
2. J. W. Gofman, *Radiation and Human Health* (Sierra Club Press, San Francisco, 1981).
3. H. Caldicott, *Nuclear Madness* (Bantam, New York, 1981).

CHAPTER 2

1. P. Nulty, "Get Ready for Power Brownouts," *Fortune,* June 5, 1989.
2. U.S. Council for Energy Awareness, "USCEA 1988 International Reactor Survey" (1989).
3. Dept. of Energy/Energy Information Agency, Monthly Energy Review (March 1989).

CHAPTER 3

1. National Academy of Sciences Committee on Carbon Dioxide Assessment, "Changing Climate," National Academy Press, 1983.
2. "The Greenhouse Effect: How It Can Change Our Lives," *EPA Journal*, Vol. 15, No. 1 (1989).
3. National Academy of Sciences, *Acid Deposition: Long Term Trends* (National Academy Press, 1986).
4. *Science News, 136*, 56 (July 22, 1989).
5. U.S. Environmental Protection Agency, "Trends in the Quality of the Nation's Air" (1988).
6. R. Wilson, S. D. Colome, J. D. Spengler, and D. G. Wilson, *Health Effects of Fossil Fuel Burning* (Ballinger Publishing Co., Cambridge, 1980).
7. W. Winkelstein *et al.* "The Relationship of Air Pollution and Economic Status to Total Mortality and Selected Respiratory System Mortality for Men," *Archives of Environmental Health 14,* 162 (1967).
8. For list of references, see B. L. Cohen, *Before It's Too Late* (Plenum Publishing, 1983) p. 114, ref. 30.
9. For list of references, see B. L. Cohen, *Before It's Too Late* (Plenum Publishing, 1983) p. 115, ref. 31.
10. U.S. Environmental Protection Agency, "Air Quality Criteria for Particulate Matter and Sulfur Oxides" (1981); "Air Quality Criteria for Oxides of Nitrogen" (1980); "Health Assessment Document for Polycyclic Organic Matter," EPA 600/9-7-008 (1979); "Health Assessment Document for Arsenic" 1980; "Air Quality Criteria for Lead" (1977).
11. H. Ozkaynak and J. C. Spengler, "Analysis of Health Effects Resulting from Population Exposure to Acid Precipitation," *Environmental Health Perspective, 63,* 45 (1985). This paper concludes that 6% of all deaths in the United States are due to air pollution. However, in private conversations with the authors, they now consider 2-5% to be a better estimate. If the U.S. population were in age equilibrium, there would be about 3 million deaths per year, which means that 60,000-150,000 deaths per year would be from air pollution. For simplicity we take a single intermediate value, 100,000.
12. J. M. Fowler, *Energy and the Environment* (McGraw-Hill, New York, 1984).
13. B. L. Cohen, "Perspective on Occupational Mortality Risks," *Health Phys. 40,* 703 (1981).
14. "The Politics of Climate," *EPRI Journal,* June 1988, p. 4.

CHAPTER 4

1. S. Rothman and S. R. Lichter, "The Nuclear Energy Debate: Scientists, the Media, and the Public," *Public Opinion,* August 1982, p. 47.

CHAPTER 5

1. S. Novick, *The Careless Atom* (Dell Publishing, New York, 1969), p. 105.
2. See Chapter 5 Appendix.
3. National Council on Radiation Protection and Measurements (NCRP), "Natural Background Radiation in the United States," NCRP Report No. 45 (1975).
4. "Report of the President's Commission on The Accident at Three Mile Island," Washington, D.C. (1979); "Three Mile Island, A Report to the Commissioners and to the Public," Nuclear Regulatory Commission Special Inquiry Group; Ad Hoc Interagency Dose Assessment Group, "Population Dose and Health Impact of the Accident at the Three Mile Island Nuclear Station," Nuclear Regulatory Commission Document NUREG-0558 (1979). Early assessment gave an average dose of 1.7 mrem, but later revisions reduced this to 1.2 mrem.
5. James Hardin (Kentucky Department of Human Resources), private communication. He was in charge of environmental monitoring in the area.
6. *Philadelphia Evening Bulletin* (May 6, 7, 8, 1979).
7. Private communication with health physicists from the Ginna plant.
8. L. Garfinkel, C. E. Poindexter, and E. Silverberg, "Cancer Statistics—1980," American Cancer Society (1981).
9. United Nations Scientific Committee on Effects of Atomic Radiation (UNSCEAR), "Sources and Effects of Ionizing Radiation," United Nations, New York (1977).
10. National Council on Radiation Protection and Measurements, Radiation Exposure from Consumer Products and Miscellaneous Sources, NCRP Report No. 56, Washington, D.C. (1977).
11. National Academy of Sciences Committee on Biological Effects of Ionizing Radiation, "Health Effects of Exposures to Low Levels of Ionizing Radiation" (BEIR-V), Washington, D.C. (1990). Note discussion in Chapter 5 on how we correct for low dose rate.
12. United Nations Scientific Committee on Effects of Atomic Radiation (UNSCEAR), "Sources, Effects, and Risks of Ionizing Radiation," United Nations, New York (1988).
13. B. L. Cohen and I. S. Lee, "A Catalog of Risks," *Health Physics, 36*, 707 (1979).
14. National Academy of Sciences Committee on Biological Effects of Ionizing Radiation, "Health Risks of Radon and Other Internally Deposited Alpha-Emitters," Washington, D.C., 1988.
15. U.S. National Council on Radiation Protection and Measurements (NCRP), "Evaluation of Occupational and Environmental Exposures to Radon and Radon Daughters in the United States," NCRP Report No. 78 (1984).
16. International Commission on Radiological Protection (ICRP), *Risk from Indoor Exposure of Radon Daughters,* ICRP Publication No. 50 (Pergamon Press, Oxford, 1987).
17. National Academy of Sciences Committee on Biological Effects of Ionizing Radiation (BEIR), "The Effects on Populations of Exposure to Low Levels of Ionizing Radiation," Washington, D.C. (1980).

⁎ 18. B. L. Cohen, "Alternatives to the BEIR Relative Risk Model for Explaining A-Bomb Survivor Cancer Mortality," *Health Physics* 52, 55 (1987).

19. U.S. National Council on Radiation Protection and Measurements (NCRP), "Influence of Dose and Its Distribution in Time on Dose-Response Relationships for Low LET Radiation," NCRP Report No. 64 (1980).

20. R. Garrison, U.S. Department of Energy, private communication, on transport accidents. Estimates for others from various sources of information.

21. C. C. Lushbaugh, S. A. Fry, C. F. Hubner and R. C. Ricks, "Total-Body Irradiation: A Historical Review and Follow-up," in C. F. Hubner and S. A. Fry (eds.), *The Medical Basis for Radiation Accident Preparedness* (Elsevier-North Holland, Amsterdam, 1980).

22. T. F. Mancuso, A. Stewart, and G. Kneale, *Health Physics, 33*, 369 (1977).

23. E. S. Gilbert, Battelle Pacific Northwest Laboratory Document PNL-SA-6341 (1978); G.B. Hutchison, B. MacMahon, S. Jablon, C. E. Land, *Health Physics,* 37, 207 (1979); U.S. General Accounting Office, "Problems in Assessing the Cancer Risks of Low-Level Ionizing Radiation Exposure," Report EMD-81, Washington, D.C. (1981); J. A. Reissland, "An assessment of the Mancuso study," Publication NRPB-79, U.K. National Radiological Protection Board, Didcot, Berk. (1978); T. W. Anderson, *Health Physics,* 35, 743 (1978); A. Brodsky, testimony before the Subcommittee on Health and the Environment, U.S. House of Representatives, Washington, D.C., 8 February 1978; B. L. Cohen, *Health Physics,* 35, 582 (1978); S. M. Gertz, ibid., 35, 723 (1978); E. S. Gilbert, "Methods of Analyzing Mortality of Workers Exposed to Low Levels of Ionizing Radiation," Report BNWL-SA-634, Battelle Pacific Northwest Laboratory, Richland, Washington (May 1977); E. Gilbert and S. Marks, *Health Physics,* 37, 791 (1979); ibid., 40, 125 (1981); J. W. Gofman, ibid, 37, 617 (1979); D. J. Kleitman, "Critique of Mancuso-Stewart-Kneale Report" (prepared for the U.S. Nuclear Regulatory Commission, Washington, D.C., 1978); S. Marks, E. S. Gilbert, and B. D. Breitenstein, "Cancer mortality in Hanford workers," Document IAEA-SM-224, International Atomic Energy Agency, Vienna (1978); R. Mole, *Lancet, i,* 582 (1978); "Staff Committee Report of November 1976," Nuclear Regulatory Commission, Washington, D.C. (1976); "Staff Committee Report of May 1978," Nuclear Regulatory Commission, Washington, D.C. (1978); "The Windscale Inquiry," Her Majesty's Stationery Office, London (1978); D. Rubenstein, "Report to the U.S. Nuclear Regulatory Commission," Nuclear Regulatory Commission, Washington, D.C. (1978); L. A. Sagan, "Low-Level Radiation Effects: The Mancuso Study," Electric Power Research Institute, Palo Alto, California (1978); B. S. Sanders, *Health Physics,* 34, 521 (1978); F. W. Spiers, ibid., *37,* 784 (1979); G. W. C. Tait, ibid., *37,* 251 (1979).

24. The Media Institute, "Television Evening News Covers Nuclear Energy," Washington, D.C. (1979).

25. B. L. Cohen, "Perspective on Genetic Effects of Radiation," *Health Physics, 46,* 1113 (1984).

26. E. B. Hook, "Rates of Chromosome Abnormalities at Different Maternal Ages," *Obstetrics and Gynecology, 58,* 282 (1981).

27. J. M. Friedman, "Genetic Disease in the Offspring of Older Fathers," *Obstetrics and Gynecology, 57,* 745 (1981).

28. International Commission on Radiological Protection (ICRP), *Recommendations of the International Commission on Radiological Protection,* ICRP Publication No. 26 (Pergamon Press, Oxford, 1977).

29. R. J. Lewis (ed.), "Registry of Toxic Effects of Chemical Substances," U.S. Public Health Service, November (1981) (available by computer access). L. Fishbein, in *Chemical Mutagens,* Vol. 4, A. Hollaender (ed.) (Plenum, New York, 1976) pp. 219ff.

30. K. Sax and H. J. Sax, "Radiomimetric Beverages, Drugs, and Mutagens," *Proceedings of the National Academy of Sciences,* 55, 1431 (1966).

31. L. Ehrenberg, G. von Ehrenstein, and A. Hedgram, "Gonad Temperature and Spontaneous Mutation Rate in Man," *Nature,* December 2, 1433 (1957).

32. U.S. Department of HEW, "Antenatal Diagnosis," National Institutes of Health Publication No. 79-1973 (1979).

33. G. W. Beebe, H. Kato, and C. E. Land, "Mortality Experience of Atomic Bomb Survivors 1950-1974," Radiation Effects Research Foundation Technical Report RERF TR 1-77 (1977). The data for Hiroshima and Nagasaki were added here.

34. National Council on Radiation Protection and Measurements (NCRP), "Review of NCRP Radiation Dose Limit for Embryo and Fetus in Occupationally Exposed Women," NCRP Report No. 53 (1977).

CHAPTER 6

1. "Report of the President's Commission on The Accident at Three Mile Island," J. B. Kemeny (Chairman), Washington, D.C., October (1979).

2. M. Rogovin (Director), "Three Mile Island, A Report to the Commissioners and to the Public," Washington, D.C., January (1980).

3. J. R. Lamarsh, *Introduction to Nuclear Engineering* (Addison-Wesley, Reading, Massachusetts, 1975); S. Glasstone and W. H. Jordan, *Nuclear Power and its Environmental Effects* (American Nuclear Society, La Grange Park, Ill., 1980).

4. G. Masche, "Systems Summary of a Westinghouse Pressurized Water Reactor Nuclear Power Plant," Westinghouse Electric Co. (1971).

5. S. Hoffman and T. Moore, "General Description of a Boiling Water Reactor," General Electric Co. (1976).

6. "American National Standards for Decay Heat Power in Light Water Reactors," American National Standards Institute ANSI/ANS-5. 1-1979.

7. "Reactor Safety Study," Nuclear Regulatory Commission Document WASH-1400, NUREG 75/014 (1975).

8. "Analysis of Three Mile Island Unit 2 Accident," Nuclear Safety Analysis Center Report NSAC-1, Palo Alto, California (July 1979); "Nuclear Accident and

Recovery at Three Mile Island," Senate Committee on Environment and Public Works, Serial No. 96-14 (July 1980); "Investigation of the March 28, 1979 Three Mile Island Accident," U.S. Nuclear Regulatory Commission Document NUREG-0600 (August 1979); Three Mile Island: The Most Studied Nuclear Accident in History," Report to the Congress by the Comptroller-General, U.S. General Accounting Office Report EMD-80-l09 (September 9, 1980).

9. Report of the Special Review Group, "Lessons Learned from Three Mile Island," U.S. Nuclear Regulatory Commission Document NUREG-0616 (December 1979).

10. "The Safety of Nuclear Power Plants and Related Facilities," U.S. AEC Report WASH-1250 (July 1973).

11. B. L. Cohen, "Physics of the Reactor Melt-down Accident," *Nuclear Science and Engineering, 80,* 47 (1982).

12. R. Gillette, "Nuclear Reactor Safety," *Science, 176,* 492 (5 May 1972); *177,* 771, 1 Sept; *177,* 867, 8 Sept; *177,* 970, 15 Sept; *177,* 1080 (22 September 1972).

13. I. Forbes, J. MacKenzie, D. F. Ford, and H. W. Kendall, "Cooling Water," *Environment* (January 1972), p. 40; D. F. Ford and H. W. Kendall, "Nuclear Safety," *Environment,* (September 1972).

14. M. L. Russel, C. W. Solbrig, and G. D. McPherson, "LOFT Contribution to Nuclear Power Reactor Safety and PWR Fuel Behavior," *Proceedings, the American Power Conference, 41,* 196 (1979); J. C. Lin, "Post Test-Analysis of LOFT Loss of Coolant Experiment L2-3," EG&G Idaho Report EGG-LOFT5075 (1980); J. P. Adams, "Quick Look Report on LOFT Nuclear Experiment L2-5," EG&G Idaho Report EGG-LOFT-5921 (1982).

15. W. Marshall (Chairman of Study Group), "An Assessment of the Integrity of PWR Pressure Vessels," U.K. Atomic Energy Authority (March 1982).

16. W. A. Carbiener *et al.,* "Physical Processes in Reactor Meltdown Accidents," Appendix VIII to Nuclear Regulatory Commission Document WASH-1400 (1975).

17. "Severe Accident Risks: An Assessment for Five U.S. Nuclear Power Plants," U.S. Nuclear Regulatory Commission Doc. NUREG-1150 (1989).

18. W. R. Butler, C. G. Tinkler, and L. S. Rubinstein, "Regulatory Perspective on Hydrogen Control for LWR Plants, "Workshop on Impact of Hydrogen on Water Reactor Safety, Albuquerque, New Mexico (January 1981); W. R. Butler and C. G. Tinkler, "Regulatory Perspective on Hydrogen Control for Degraded Core Accidents," Second International Workshop on the Impact of Hydrogen on Water Reactor Safety, Albuquerque, New Mexico (1982).

19. "Hydrogen Control for Sequoyah Nuclear Plant," Nuclear Regulatory Commission Document dated August 13, 1980.

20. "Proposed Interim Hydrogen Control Requirements for Small Containments," Memorandum from H. Denton to The NRC commissioners dated February 22, 1980, NRC Document SECY-80-l07.

21. Union of Concerned Scientists, "The Risks of Nuclear Power Reactors," Cambridge, Massachusetts (1977).

22. H. W. Lewis (Chairman), "Risk Assessment Review Group Report to the U.S. Nuclear Regulatory Commission," NUREG/CR-400 (1978).
23. F. J. Rahn and M. Levenson, "Radioactivity Releases Following Class-9 Reactor Accidents," Health Physics Society, Las Vegas, Nevada (June 1982); C. D. Wilkinson, "NSAC Workshop on Reactor Accident Iodine Release," Palo Alto, California (July 1980); H. A. Morewitz, "Fission Product and Aerosol Behavior Following Degraded Core Accidents," *Nuclear Technology, 53,* 120 (1981).
24. R. Wilson, S. D. Colome, J. D. Spengler, and D. G. Wilson, *Health Effects of Fossil Fuel Burning.* (Ballinger, Cambridge, Massachusetts, 1980).
25. International Symposium on Areas of High Natural Radioactivity, Academy of Sciences of Brazil (June 1975).
26. W. Ramsay, *The Unpaid Costs of Electrical Energy* (Johns Hopkins University Press, Baltimore, Maryland, 1979).
27. "Draft NRC Evaluation of Pressurized Thermal Shock" (September 13, 1982); T. A. Meyer, "Summary Report on Reactor Vessel Integrity for Westinghouse Operating Plants," Westinghouse Electric Corp. Report WCAP-10019 (December 1981); "Summary of Evaluations Related to Reactor Vessel Integrity Performed for the Westinghouse Owner's Group," Westinghouse Electric Corp., Nuclear Technology Division (May 1982).
28. M. L. Wald, "Steel Turned Brittle by Radiation Called a Peril at 13 Nuclear Plants," *New York Times* (September 27, 1981).
29. R. Immel, "Stress Corrosion Cracking," *EPRI Journal* (November 1981).
30. "Steam Generator Tube Experience," U.S. Nuclear Regulatory Commission Document NUREG-0886 (1982).
31. "Report on the January 25, 1982 Steam Generator Tube Rupture at the R.E. Ginna Nuclear Power Plant," NRC Document NUREG-0909 (April 1982).

Chapter 7

1. B. L. Cohen, "The Nuclear Reactor Accident at Chernobyl, USSR," *American Journal of Physics, 55,* 1076 (1987).
2. Soviet Report on the Chernobyl Accident, English Translation. U.S. Dept. of Energy, Washington, D.C. (August 1986).
3. United Nations Scientific Committee on Effects of Atomic Radiation, "Sources, Effects, and Risks of Ionizing Radiation," New York (1988).

Chapter 8

1. B. L. Cohen and I. S. Lee, "A Catalog of Risks," *Health Physics, 36,* 707 (1979). Numerous references are given to original sources of information.
2. Chap. 3, ref. 11.

3. U.S. National Academy of Sciences Committee on Biological Effects of Ionizing Radiation, "Health Risks of Radon and Other Internally Deposited Alpha Emitters," (BEIR IV), Washington, D.C. (1988).

4. B. L. Cohen, "Perspective on Occupational Risks," *Health Physics, 40,* 703(1981).

5. National Center for Health Statistics, *Monthly Vital Statistics Reports, 36,* 5(Supplement)(1987).

6. B. L. Cohen, "Catalog of Risks Extended," to be published by Task Force on Risk Analysis, American Association of Engineering Societies (1990).

7. World Resources Institute, *World Resources 1987* (Basic Books, New York, 1987).

8. C. L. Comar and L. A. Sagan, "Health Effects of Energy Production and Conversion," *Annual Review of Energy, 1,* 581 (1976); L. B. Lave and L. C. Freeburg, "Health Effects of Electricity Generation from Coal, Oil, and Nuclear Fuel," *Nuclear Safety, 14*(5), 409, (1973); S. M. Barrager, B. R. Judd, and D. W. North, "The Economic and Social Costs of Coal and Nuclear Electric Generation," Stanford Research Institute Report (March 1976); Nuclear Energy Policy Study Group, *Nuclear Power—Issues and Choices* (Ballinger, Cambridge, Massachusetts, 1977); National Academy of Sciences Committee on Nuclear and Alternative Energy Systems, *Energy in Transition, 1985-2010* (W. H. Freeman, San Francisco, 1980); American Medical Association Council on Scientific Affairs, "Health Evaluation of Energy Generating Sources," *Journal of the American Medical Association, 240,* 2193 (1978); H. Inhaber, "Risk of Energy Production," Atomic Energy Control Board Report AECB-1119, Ottawa (1978); R. L. Gotchy, "Health Effects Attributable to Coal and Nuclear Fuel Cycle Alternatives," U.S. Nuclear Regulatory Commission Document NUREG-0332 (1977); D. J. Rose, P. W. Walsh, and L. L. Leskovjan, "Nuclear Power—Compared to What?" *American Scientist, 64,* 291 (1976); Union of Concerned Scientists, "The Risks of Nuclear Power Reactors," H. Kendall (Director), Cambridge, Massachusetts (1977); Science Advisory Office, State of Maryland, "Coal and Nuclear Power" (1980); Norwegian Ministry of Oil and Energy, "Nuclear Power and Safety" (1978); Ohio River Basin Energy Study (EPA), "Impacts on Human Health from Coal and Nuclear Fuel Cycles" (July 1980); United Kingdom Health and Safety Executive, "Comparative Risks of Electricity Production Systems" (1980); Maryland Power Plant Siting Program, "Power Plant Cumulative Environmental Impact Report," PPSP-CEIR-1 (1975); W. Ramsay, *Unpaid Costs of Electrical Energy* (Johns Hopkins Univ. Press, Baltimore, Maryland, 1979); Legislative Office of Science Advisor, State of Michigan, "Coal and Nuclear Power" (1980); B. L. Cohen, *American Scientist, 64,* 291 (1976); R. Wilson and W. J. Jones, *Energy, Ecology, and the Environment* (Academic Press, New York, 1974); H. Fischer *et al.,* "Comparative Effects of Different Energy Technologies," Brookhaven National Lab Report BNL 51491 (September 1981); D. K. Myers and H. B. Newcombe, "Health Effects of Energy Development," Atomic Energy of Canada Ltd. Report AECL-6678 (1980); B.

L. Cohen, *Before It's Too Late* (Plenum, New York, 1983); American Medical Association Council on Scientific Affairs, "Medical Perspective on Nuclear Power" (1989).

9. B. L. Cohen, Long-Term Consequences of the Linear-No Threshold Dose-Response Relationship for Chemical Carcinogens, *Risk Analysis 1*, 267 (1981).
10. B. L. Cohen, "Society's Valuation of Life Saving," *Health Physics, 38*, 33(1980). Numerous references to original sources of information are included.
11. B. L. Cohen, "Reducing the Hazards of Nuclear Power: Insanity in Action," *Physics and Society, 16*, 3, 29(1987).
12. B. L. Cohen, "Cost Effectiveness of Reducing Radon Levels in Homes," *Journal of Nuclear Medicine, 29*, 268(1988).

CHAPTER 9

1. United Engineers and Constructors, Inc., "Phase IX Update (1987) Report for the Energy Economic Data Base Program" (July 1988). Also, "Phase × Update Report" (September 1989).
2. A. Reynolds, "Cost of Coal vs. Nuclear in Electric Power Generation," U.S. Energy Information Administration Document (1982).
3. W. W. Brandfon, "The Economics of Nuclear Power," American Ceramic Society, Cincinnati (1982).
4. I. Spiewak and D. F. Cope, "Overview Paper on Nuclear Power," Oak Ridge National Laboratory Report ORNL/TM-7425.
5. J. H. Crowley, "Nuclear Energy—What's Next," Atomic-Industrial Forum Workshop on the Electric Imperative, Monterey, CA (1981).
6. Long Island Lighting Company, "The Shoreham Nuclear Power Plant: An Overview" (1982).
7. M. R. Copulos, *Confrontation at Seabrook*, The Heritage Foundation (1978).

CHAPTER 10

1. J. J. Taylor, "Improved and Safer Nuclear Power" *Science, 244*, 318(April 21, 1989).
2. J. J. Taylor, K. E. Stahlkopf, D. M. Noble, and G. J. Dau, "LWR Development in the U.S.A.," *Nuclear Engineering and Design 109*, 19 (1988).
3. J. Catron, "New Interest in Passive Reactor Designs," *EPRI Journal, 14*, 3, 4 (April 1989).
4. R. Livingston, "The Next Generation," Nuclear Industry (July 1988).
5. K. E. Stahlkopf, J. C. DeVine, and W. R. Sugnet, "U.S. ALWR Programme Sets Out Utility Requirements for the Future," *Nuclear Engineering International* (November 1988), p. 16.

6. R. Vijuk and H. Bruschi, "AP-600 Offers a Simpler Way to Greater Safety, Operability, and Maintainability," *Nuclear Engineering International* (November 1988), p. 22.

7. K. Hannerz, "Applying PIUS to Power Generation," *Nuclear Engineering International* (December 1983).

8. K. Hannerz, "Making Progress on PIUS Design and Verification," *Nuclear Engineering International* (November 1988), p. 29.

9. "Nuclear Power: The New Generation," *IAEA* (International Atomic Energy Agency) *Bulletin, 331,* No. 3(entire issue) (1989).

10. Edison Electric Institute, Statistical Yearbook of the Electrical Utility Industry, 1981.

11. U.S. Energy Information Agency, "Historical Plant Cost and Annual Production Expenses for Selected Electricity Generating Plants," DOE/EIA-0455 (1987).

12. U.S. Department of Energy, "Nuclear Energy Cost Data Base," DOE/NE-0095 (September 1988).

13. Study Group of the Committee on Financial Considerations, "A Comparison of the Future Costs of Nuclear and Coal-Fired Electricity, An Update," U.S. Council for Energy Awareness (1987).

14. *Nuclear News* (July 1982), p. 48.

CHAPTER 11

1. R. Wilson, S. D. Colome, J. D. Spengler, and D. G. Wilson, *Health Effects of Fossil Fuel Burning* (Ballinger, Cambridge, Massachusetts, 1980).

2. A. G. Croff and C. W. Alexander, "Decay Characteristics of Once-Through LWR and LMFBR Spent Fuels, High Level Wastes, and Fuel Assembly Structural Materials Wastes," Oak Ridge National Laboratory Report ORNL/TM-7431 (1980).

3. Reference 1 gives the risk from sulfur dioxide (SO2) as 3.5×10^{-5}/year for 1 microgram SO_2 per meter^3 of air. An average person inhales 7,000 meters^3 of air per year, so this corresponds to inhaling 7,000 micrograms, or 0.007g, of SO_2. The deaths per gram of SO_2 are then $(3.5 \times 10^{-5}/0.007 =) 0.005$. An average coal-burning plant produces 3×10^8g of SO_2 per day; this could then cause $(3 \times 10^8 \times 0.005 =) 1.5$ million deaths if it were all inhaled by people.

4. See Figure 1 (Chapter 11), which gives the effects from eating all the waste produced in one year. This number must be divided by the number of days per year to obtain the effects from one day.

5. B. L. Cohen, "Ocean Dumping of Radioactive Waste," *Nuclear Technology, 47,* 163 (1980). Some of the numbers quoted in that paper have been changed due to later data, but these are incorporated into the results quoted here.

6. "Air Quality Criteria for Particulate Matter and Sulfur Oxides," U.S. Environmental Protection Agency (February 1981); "Air Quality Criteria for Oxides of Nitrogen," U.S. Environmental Protection Agency Report (June 1980).

7. U.S. Environmental Protection Agency, Health Assessment Document for Polycyclic Organic Matter, Report EPA-600/9-79-008 (1979).
8. B. L. Cohen, "Risk Analysis of Buried Waste from Electricity Generation," *American Journal of Physics, 54,* 38 (1986). This contains references to the original papers on which many of the analyses here are based.
9. These are reviewed by C. M. Koplik, M. F. Kaplan, and B. Ross, "The Safety of Repositories for Highly Radioactive Waste," *Reviews of Modern Physics, 54,* 269 (1982); Interagency Review Group, Report to the President, U.S. Department of Energy Report TID-29442 (1978).
10. B. L. Cohen, "Probability for Human Intake of an Atom Randomly Released into the Ground, Rivers, Oceans, and Air," *Health Physics, 47,* 281 (1984).
11. B. L. Cohen, "High Level Waste from Light Water Reactors," *Reviews of Modern Physics, 49,* 1 (1977); The plans for waste burial have changed somewhat since that time, including a threefold dilution of the waste in glass. This is taken into account in the numbers given here.
12. J. A. Reuppen, M. A. Molecke, and R. S. Glass, "Titanium Utilization in Long Term Nuclear Waste Storage," Sandia National Lab Report, SAND81-2466 (1981).
13. M. F. Kaplan, "Characterization of Weathered Glass by Analyzing Ancient Artifacts," in *Scientific Basis for Nuclear Waste,* C. J. M. Northrup (ed.) (Plenum, New York, 1980).
14. K. J. Schneider and A. M. Platt, "High Level Radioactive Waste Management Alternatives," Battelle Northwest Laboratory Report BNWL-1900 (1974).
15. B. L. Cohen, "Analysis, Critique, and Re-evaluation of High Level Waste Repository Water Intrusion Scenario Studies," *Nuclear Technology, 48,* 63 (1980).
16. B. L. Cohen, "Discounting in Assessment of Future Radiation Effects," *Health Physics, 45,* 687 (1983).
17. U.S. Department of Energy, Office of Civilian Radioactive Waste Management, "What is Tuff?" (1988); "What Happens During Site Characterization?" (1988); "Why Yucca Mountain?" (1988).
18. G. Russ, "Notes from Underground," *Nuclear Industry* (Spring 1989), p. 13.
19. B. L. Cohen, "A Simple Probabilistic Risk Analysis for High Level Waste Repositories," *Nuclear Technology, 68,* 73 (1985).

CHAPTER 12

1. National Academy of Sciences Committee on Biological Effects of Ionizing Radiation, "Health Risks of Radon" (BEIR-IV), 1988.
2. B. L. Cohen, "Risk Analysis of Buried Wastes from Electricity Generation," *American Journal of Physics, 54,* 38 (1986). This is a review article which gives references to the original papers.
3. Code of Federal Regulations, Title 10, Part 50, Appendix I; U.S. Nuclear Regulatory Commission, Regulatory Guides 1.109-1.113(1976). These specify expendi-

ture of $1,000/man-rem. Dividing this by the risk given in Chapter 5, 260×10^{-6} per man-rem, gives $4 million per cancer death averted.

4. United Nations Scientific Committee on Effects of Atomic Radiation, "Sources, Effects, and Risks of Ionizing Radiation," United Nations, New York (1988).

5. Code of Federal Regulations, Title 10, Part 61.

6. U.S. Energy Research and Development Administration (ERDA), "Alternatives for Long Term Management of Defense High-Level Radioactive Waste," Document ERDA-77-42/1 (1977).

7. L. R. Brown, "World Food Resources and Population," Population Reference Bureau (1981).

8. B. L. Cohen, "The Situation at West Valley," *Public Utilities Fortnightly* (September 27, 1979), p.26.

9. Western New York Nuclear Service Center Study, U.S. Department of Energy Report TID-28905-2 (1980).

10. R. M. Jefferson and H. R. Yoshimura, "Crash Testing of Nuclear Fuel Shipping Containers," Sandia National Lab Report SAND 77-1462 (1978).

11. Nuclear Regulatory Commission, "Final Environmental Statement on the Transportation of Radioactive Material by Air and other Modes," Document NUREG-1070 (1977).

12. R. P. Sandoval and G. J. Newton, "A Safety Assessment of Spent Fuel Transportation Through Urban Regions," Sandia National Laboratory Report SAND 81-2147 (1981); R. P. Sandoval, "Safety Assessment of Spent Fuel Transport in an Extreme Environment," *Nuclear and Chemical Waste Management, 3,* 5 (1982).

13. Z. A. Medvedev, *New Scientist, 72,* 264 (1976); *72,* 692 (1976); *74,* 761 (1977); *76,* 352 (1977).

14. J. R. Trabalka, L. D. Eyman, and S. I. Auerbach, "Analysis of the 1957-58 Soviet Nuclear Accident," *Science, 209,* 345 (18 July 1980).

15. D. M. Soran, and D. B. Stillman, "An Analysis of the Alleged Kyshtym Disaster," Los Alamos National Lab Report LA-9217-MS(1981).

16. International Atomic Energy Agency Information Circular, "Report on a Radiological Accident in the Southern Urals on 29 September 1957," INFCIRC/368 (28 July 1989).

CHAPTER 13

1. F. R. Best and M. J. Driscoll, *Transactions of the American Nuclear Society, 34,* 380 (1980).

2. B. L. Cohen, Breeder Reactors—A Renewable Energy Source, *American Journal of Physics, 51,* 75 (1983).

3. R. Avery and H. A. Bethe, "Breeder Reactors: The Next Generation," in *Nuclear Power: Both Sides,* M. Kaku and J. Trainer (eds.) (Norton, New York, 1982).

4. T. G. Ayers *et al.,* LMFBR Program Review, U.S. Energy Research and Develop-

ment Administration (1978); Report of the Task Forces to the LMFBR Review Steering Committee, Energy Research and Development Administration (April 6, 1977).

5. Joint Committee on Atomic Energy, Review of National Breeder Reactor Program (January 1976).

6. R. Wilson, "Report on the Safety of a Liquid Metal Fast Breeder Reactor," Electric Power Research Institute (Palo Alto, CA, 1976).

7. J. B. Yasinsky (ed.), "Position Papers on Major Issues Associated with the Liquid Metal Fast Breeder Reactor," Westinghouse Electric Corporation (Madison, Pennsylvania) 1978.

8. Milton R. Benjamin, *Washington Post* (July 20, 1982).

9. Atomic-Industrial Forum, "Light Water Reactor Fuel Cycles—An Economic Comparison of the Recycle and Throw-away Alternatives" (February 1981). This presents analyses by six different groups.

10. Robert Lesch, "World Reprocessing Facilities," *Worldwide Nuclear Power* (January 1982).

11. Shelby T. Brewer, "Letter to Recipients of Worldwide Nuclear Power" (dated March 4, 1982).

12. W. Meyer, S. K. Loyalka, W. E. Nelson, and R. W. Williams, "The Homemade Bomb Syndrome," *Nuclear Safety, 18,* 427 (1977).

13. C. Starr and E. Zebroski, "Nuclear Power and Weapons Proliferation," American Power Conference (April 1977).

14. J. McPhee, *The Curve of Binding Energy* (Ballantine Books, New York, 1975).

15. U.S. Nuclear Regulatory Commission, "Safeguarding a Domestic Mixed Oxide Industry Against a Hypothetical Subnational Threat," NUREG-0414 (May 1978).

16. U.S. Nuclear Regulatory Commission, "Regulatory Guide 5.55; Standard Format and Content of Safeguards Contingency Plans for Fuel Cycle Facilities" (1978); also "Regulatory Guide 5.54: Standard Format and Content of Safeguards Contingency Plans for Nuclear Power Plants" (1978).

17. U.S. Nuclear Regulatory Commission, "Regulatory Guide 5.56: Standard Format and Content of Safeguards Contingency Plans for Transportation" (1978).

18. M. Willrich and T. B. Taylor, *Nuclear Theft: Risks and Safeguards* (Ballinger, Cambridge, Massachusetts, 1974).

19. F. H. Schmidt and D. Bodansky, *The Energy Controversy: The Fight over Nuclear Power* (Albion Press, San Francisco, 1976).

20. B. L. Cohen, "Plutonium—How Great is the Terrorist Threat," *Nuclear Engineering International* (February 1977). The quote from Kinderman is given there. It was taken from a book, but I cannot recall the name of the latter.

21. U.S. Energy Research and Development Administration (ERDA) Safeguards Program, Background Statement (March 10, 1975).

22. R. Nader, speech at Lafayette College (Spring 1975).

23. International Commission on Radiological Protection (ICRP), Task Group on Lung Dynamics, "Deposition and Retention Models for Internal Dosimetry of the Human Respiratory Tract," *Health Physics, 12,* 173 (1966).

24. National Academy of Sciences Committee on Biological Effects of Ionizing Radiation (BEIR), "Effects on Populations of Exposure to Low Levels of Ionizing Radiation" (1980).

25. United Nations Scientific Committee on Effects of Atomic Radiation (UNSCEAR), "Sources and Effects of Ionizing Radiation" (1977).

26. B. L. Cohen, "Hazards from Plutonium Toxicity," *Health Physics, 32,* 359 (1977).

27. The Medical Research Council, *The Toxicity of Plutonium* (Her Majesty's Stationery Office, London, 1975); C. W. Mays, "Discussion of Plutonium Toxicity," in R. G. Sachs (ed.), *National Energy Issues—How Do We Decide* (Ballinger, Cambridge, Massachusetts, 1980).

28. W. J. Bair, "Toxicity of Plutonium," *Advances in Radiation Biology,* Vol. 4, p. 225 (1974).

29. J. H. Rothchild, *Tomorrow's Weapons* (McGraw-Hill, New York, 1964).

30. J. F. Park, W. J. Bair, and R. H. Busch, "Progress in Beagle Dog Studies with Transuranium Elements at Battelle-Northwest," *Health Physics, 22,* 803 (1972).

31. A. R. Tamplin and T. B. Cochran, "Radiation Standards for Hot Particles," Natural Resources Defense Council Report (1974). Also, "Petition to Amend Radiation Protection Standards as They Apply to Hot Particles," submitted to EPA and AEC (February 1974).

32. National Academy of Sciences, "Health Effects of Alpha Emitting Particles in the Respiratory Tract," Environmental Protection Agency Document EPA 520/4-76-013 (1976); National Council on Radiation Protection and Measurements (NCRP), "Alpha Emitting Particles in Lungs," NCRP Report No. 46 (1975); United Kingdom National Radiological Protection Board, Report R-29 and Bulletin No. 8 (1974); W. J. Bair, C. R. Richmond, and B. W. Wachholz, "A Radiological Assessment of the Spatial Distribution of Dose from Plutonium," U.S Atomic Energy Commission Report WASH-1320 (1974); see also The Medical Research Council, ref. 27.

33. G. L. Voelz, "What We Have Learned About Plutonium from Human Data," *Health Physics, 29* (1975).

34. J. W. Gofman, "The Cancer Hazard from Inhaled Plutonium," Committee for Nuclear Responsibility Report CNR 1975-1, reprinted in Congressional Record-Senate 31, (July 1975), p. 14610.

35. R. W. Albert *et al., Archives of Environmental Health, 18,* 738 (1969); *Archives of Environmental Health, 30,* 361 (1975).

36. W. J. Bair, "Review of Reports by J. W. Gofman on Inhaled Plutonium," Battelle Northwest Laboratory Report BNWL-2067; C. R. Richmond, "Review of John W. Gofman's Report on Health Hazards from Inhaled Plutonium," Oak Ridge National Laboratory Report ORNL-TM-5257 (1975); J. W. Healy *et al.,* "A Brief Review of the Plutonium Lung Cancer Estimates by John W. Gofman," Los Alamos Scientific Laboratory Report LA-UR-75-1779 (1975); M. B. Snipes *et al.,* "Review of John Gofman's Papers on Lung Cancer Hazard from Inhaled Plutonium," Lovelace Foundation (Albuquerque, New Mexico) Report LF-51

UC-48 (1975); "Comments Prepared by D. Grahn," Argonne National Laboratory (1975).

37. R. Nader, *Family Health* (January 1977), p. 53.
38. U.S. Atomic Energy Commission, "Meteorology and Atomic Energy," p. 97ff (1968). This gives the calculational procedures used in ref. 26.
39. K. Stewart, in *The Resuspension of Particulate Material from Surfaces*, B. R. Fish (ed.), (Pergamon Press, New York, 1964); L. R. Anspaugh, P. L. Phelps, N. C. Kennedy, and H. C. Booth, Proceedings of the Conference on Environmental Behavior of Radionuclides Released in the Nuclear Industry, International Atomic Energy Agency (Vienna, 1975).
40. H. R. McLendon *et al.*, International Atomic Energy Agency Document IAEA-SM-199/85, p. 347 (1976) — Savannah River Plant; R. C. Dahlman, E. A. Bondietti, and L. D. Eyman, Oak Ridge National Laboratory Environmental Sciences Division Publication 870 (1976) — Oak Ridge; F. W. Wicker, Colorado State University Report COO-1156-80 (1975) — Rocky Flats, Colorado; E. M. Romney, A. Wallace, R. O. Gilbert, and J. E. Kinnear, International Atomic Energy Agency Document IAEA-SM-199/73, p. 479 (1976) — Eniwetok.
41. J. Gofman, *National Forum* (Summer, 1979).
42. B. L. Cohen, "Plutonium Containment," *Health Physics, 40,* 76 (1981).
43. U.S. Environmental Protection Agency, *Federal Register, 40,* 23420 (1975).
44. U.S. Atomic Energy Agency, "Plutonium and Other Transuranic Elements: Sources, Environmental Distribution, and Biomedical Effects," Document WASH-1359 (1974).
45. J. R. Lamarsh, *Introduction to Nuclear Engineering* (Addison-Wesley, Reading, Massachusetts, 1975).
46. International Commission on Radiological Protection (ICRP), *Report of the Task Group on Reference Man,* ICRP Publication 23 (Pergamon Press, New York, 1975).
47. International Commission on Radiological Protection (ICRP), *Recommendations of ICRP,* ICRP Publication 26 (Pergamon Press, New York, 1977).

CHAPTER 14

1. B. L. Cohen, "Cost per Million BTU of Solar Heat, Insulation, and Conventional Fuels," *American Journal of Physics, 52,* 614 (1984).
2. R. H. Annan and J. L. Stone, "The U.S. National Photovoltaics Program — Investing in Success," *Solar Cells, 26,* 135 (1989).
3. Solar Energy Research Institute, *Photovoltaics Technical Information Guide,* (Second Edition), p. 58 (1988).
4. U.S. Department of Energy, *Photovoltaic Energy Program Summary* (1988).
5. U.S. Department of Energy, *Solar Thermal Energy Program Summary* (1988).
6. U.S. Department of Energy, *Wind Energy Program Summary* (1988).
7. U.S. Department of Energy, *Energy Storage and Distribution Program Summary* (1988).

INDEX